Jochen Winkler

Dispersing Pigments and Fillers

Jochen Winkler: Dispersing Pigments and Fillers

ISBN 978-3-86630-852-7

Cover: Evonik Degussa GmbH, Essen/Germany

Jochen Winkler
Dispersing Pigments and Fillers
Hanover: Vincentz Network, 2012
EUROPEAN COATINGS TECH FILES
ISBN 3-86630-852-3
ISBN 978-3-86630-852-7

Please ask for our book catalogue
Vincentz Network, Plathnerstr. 4c, 30175 Hanover, Germany
T (202) 684-6630, F (202) 380-9129
E-mail: books@american-coatings.com, www.american-coatings.com

Layout: Vincentz Network, Hanover, Germany

ISBN 3-86630-852-3
ISBN 978-3-86630-852-7

European Coatings Tech Files

Jochen Winkler

Dispersing Pigments and Fillers

Jochen Winkler: Dispersing Pigments and Fillers

ISBN 978-3-86630-852-7

Preface

For the last 100 years or so, our everyday life has been predominated by the use polymeric materials. Tires for automobiles, for example, grant us the mobility that is a paramount precondition for our modern civilization to function as it does. More and more, polymeric materials are substituting metals or, as in the case of coatings, are delivering important contributions to the protection of valuable resources.

It is true for almost all polymeric materials that they gain their full potential of use only by the incorporation of pigments and fillers (extenders). Usually, one finds that the pigments only develop their beneficial action when they are evenly distributed within the polymeric matrices. Most pigments and fillers are produced and marketed as dry powders that "agglomerate" due to mutual attraction and therefore present themselves as larger, fairly spherical entities. The destruction of these structures by mechanical forces in polymer melts or polymer solutions, yielding a homogenous distribution of the single pigment particles is called "dispersing". In that sense, dispersing is the elemental step in the production of any composite materials, especially in the case of coatings.

In spite of the huge importance of dispersion processes in the production of composite materials, dispersing itself is still often looked upon as being more of an art rather than a fundamental, scientifically underlain technical process. The reason for this may lie in the fact that a number of separate steps take place simultaneously during dispersions. These steps are the **wetting of pigment surfaces,** the **mechanical disruption of agglomerates** and the **stabilization of the single "primary" pigment particles** obtained against renewed agglomeration, which is called "flocculation". The interrelations appear complicated and confusing to some. None of the single steps can be studied completely isolated from one another in a quantitative manner. Yet, there is profound knowledge concerning the physical and chemical background to these three steps. And, although quantitative prognoses are difficult, there are a number of perceptions and theories that lead to a sound understanding of the influencing parameters. With this knowledge, it is possible to find the reasons for possible failure on a case to case basis in a structured manner and to then find solutions to the problems quickly. In doing so, mathematical formulas are of great assistance since they enable an operator to find the influencing parameters at a glance. In the simplest case, a formula

Jochen Winkler: Dispersing Pigments and Fillers

ISBN 978-3-86630-852-7

depicts which parameters facilitate a process, which impede it and which don't play a role at all.

In order to satisfy the pretension of this book, namely to convey fundamental knowledge of dispersion processes, the basic interactions between atoms and/ or simple molecules is treated first. In that way, the wordings that are used to describe colloidal interactions are filled with meaning.

Following this are chapters in which the elemental steps of dispersions are discussed. These are explicitly:

- the wetting of pigment (filler or extender) surfaces by liquid components of a mill base
- the mechanical breakage of pigment (filler or extender) agglomerates
- the stabilization of the dispersed pigments against flocculation.

The pretension of this book is formulated in the title: "Dispersing Pigments and Fillers". It is meant to serve the experienced practitioner as well as the novice as a source for information for their everyday work.

Krefeld, January 2012
Jochen Winkler

Contents

Jochen Winkler: Dispersing Pigments and Fillers

ISBN 978-3-86630-852-7

1 Physical interactions of atoms and molecules

The periodic table arranges the elements in a clear way and, amongst other things, provides the basis for understanding chemical bonds. Covalent and ionic chemical bonds are distinguished. Depending upon the extent of the ionic portion of a chemical bond, the molecules exhibit different dipole moments and, depending upon their size, their electron shells are more or less easily polarizable. This chapter deals with the methods for determining dipole moments and polarizabilities of dielectric substances and how these two properties can be used to estimate the physical interaction energies and forces between atoms or molecules. In order to get a feeling for the relative significance of the different interaction principles, it will be demonstrated how the boiling points (as a measure for these interactions) rely on the dipole moments and polarizabilities. This chapter forms the fundament for later chapters in which the physical interactions in colloidal systems are discussed.

1.1 Periodic table of the elements

The chemical elements are assorted in the so called "periodic table of the elements", shown in Figure 1.1. Every element consists of an atomic nucleus with positively charged protons and a varying number of uncharged ("electroneutral") neutrons. The nuclei, which comprise almost the whole mass of the elements, are surrounded by the negatively charged electrons. Since the number of protons and electrons in an element are equal in number, the positive and negative charges compensate each other and the whole atom (= atomic nucleus plus electrons) carries no externally measurable net charge.

The electrons are located in the so called "orbitals". Orbitals differ in shape and size from one another. Every orbital has room for two electrons only. According to the Heisenberg uncertainty principle, it is not possible to denote both the location and the impuls (= mass · velocity) of an electron simultaneously[1]. For that reason, orbitals are to be understood as probable areas in which electrons may be encountered. Illustrations of orbitals represent exactly that, as shown in Figure 1.2.

[1] This is not easy to understand. The uncertainty principle results from quantum mechanical calculations. In these, the determination of the position of a quantum object necessarily perturbs its impuls (momentum).

Jochen Winkler: Dispersing Pigments and Fillers

ISBN 978-3-86630-852-7

←s-block p-block→

Atomic number → 1 H 2,2 ← Element; Electronegativity → 2,2

	I	II											III	IV	V	VI	VII	VIII
1	1 H 2,2																	2 He -
2	3 Li 0,98	4 Be 1,57											5 B 2,04	6 C 2,55	7 N 3,04	8 O 3,44	9 F 3,98	10 Ne -
3	11 Na 0,93	12 Mg 1,31					d-block						13 Al 1,61	14 Si 1,9	15 P 2,19	16 S 2,58	17 Cl 3,16	18 Ar -
4	19 K 0,82	20 Ca 1	21 Sc 1,36	22 Ti 1,54	23 V 1,63	24 Cr 1,66	25 Mn 1,55	26 Fe 1,83	27 Co 1,88	28 Ni 1,91	29 Cu 1,9	30 Zn 1,65	31 Ga 1,81	32 Ge 2,01	33 As 2,18	34 Se 2,55	35 Br 2,96	36 Kr -
5	37 Rb 0,82	38 Sr 0,95	39 Y 1,22	40 Zr 1,33	41 Nb 1,6	42 Mo 2,16	43 Tc 1,9	44 Ru 2,2	45 Rh 2,28	46 Pd 2,2	47 Ag 1,93	48 Cd 1,69	49 In 1,78	50 Sn 1,96	51 Sb 2,05	52 Te 2,1	53 I 2,66	54 Xe 2,6
6	55 Cs 0,79	56 Ba 0,89	57 La* 1,1	72 Hf 1,3	73 Ta 1,5	74 W 1,7	75 Re 1,9	76 Os 2,2	77 Ir 2,2	78 Pt 2,2	79 Au 2,4	80 Hg 1,9	81 Tl 1,8	82 Pb 1,8	83 Bi 1,9	84 Po 2	85 At 2,2	86 Rn -

Figure 1.1: Periodic table of the elements with atomic numbers and electronegativities listed

The periodic table consist of rows and columns. Starting from the first element, hydrogen (symbol H), and when going from left to right from one element to the next, every element is distinguished from its neighbour to the left in that it has both one additional proton and one further electron. Apart from that, a number of neutrons may accrue. When reaching the right end of a row, the next element with one additional proton and electron is the element standing at the left side of the following row. The number of electrons and protons that an atom has can therefore be taken from its "atomic number", a simple, continuous numbering of the elements in the periodic table.

A further feature of chemical elements is that they have different tendencies to attract the binding electrons in chemical bonds. The greater the electron drawing effect of an element is, the more "electronegative" it is. The electronegativity increases from left to right within a row and declines from the top to the bottom of a column. The highest electronegativities belong to the elements in the upper right corner of the periodic table. In the last column, next to the "halogens" fluorine (F), chlorine (Cl), bromine (Br), iodine (J) and astatine (As), are the "noble gases" helium (He), neon (Ne), argon (Ar), krypton (Kr) and radon (Rn). The name "noble gas" expresses both the gaseous nature of the elements and that they usually do not undergo chemical reactions (they are "noble"). This is due to the fact that the electrons are in a "low energy state", the so called "noble gas electron configuration" . This constellation is energetically favourable so that the other elements also exhibit the tendency to reach this state. They do this either by sharing electrons amongst each other, by rendering electrons to a chemical reaction partner, or by gathering electrons from them. Hence, the driving force for the generation of chemical bonds is explained by the reaction partners either sharing electrons, gaining electrons or releasing electrons to reach the noble gas

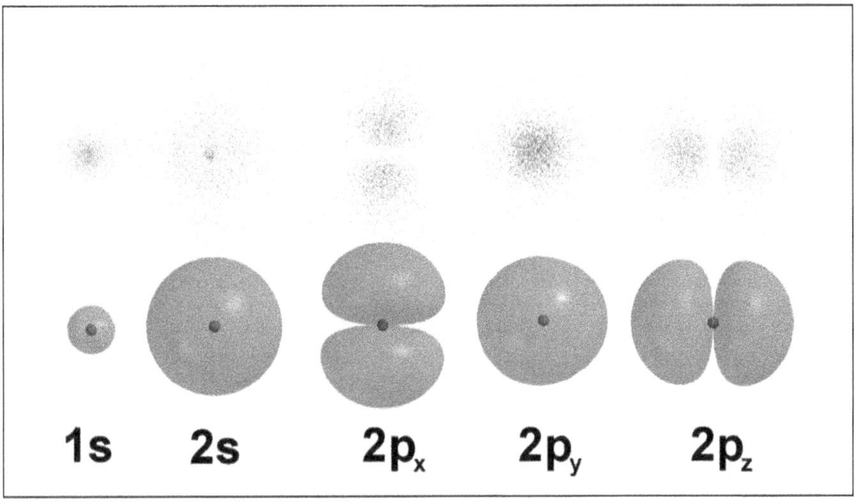

Figure 1.2: Drawings of 1s, 2s, $2p_x$, $2p_y$ and $2p_z$ orbitals as spaces with certain probabilities for the electrons to be located in. Source: Wikipedia; Search word "Orbital"; public domain

electron configuration, i.e. having eight electrons in their outer electron shell ("octet rule"). For the elements in the first (second) column of the periodic table, it is more favourable to reach this state by rendering one (two) electron(s) to its partner in a chemical reaction, whereas an element in the seventh (sixth) column preferably takes up on (two) electron(s). The electrons participating in a chemical bond are called "binding electrons".

1.1.1 Covalent and ionic bonds

When chemical elements (= atoms) react with one another, molecules are formed. Depending upon the differences in electronegativities between the reaction partners, the binding electrons are distributed more or less equally in the chemical bond. Therefore, covalent and ionic chemical bonds are distinguished. Covalent bonds are formed when partners with the same electron drawing properties react with each other. This is especially the case when two atoms of the same sort, such as two chlorine atoms, react. Covalent bonds are formed when two orbitals with one single electron each overlap, so that a binding orbital is formed which is filled with two electrons. In that case, one may say that the two reaction partner mutually "share" the electrons.

The second possibility is that one reaction partner delivers one or more electrons to the other reaction partner. In this case, ionic bonds are formed in which one of the reactants obtains a positive charge (cation) and the other a negative charge (anion). The two oppositely charged ions attract each other electrostatically, so that the atoms become chemically connected. An example for this is the reaction of lithium (Li) with fluorine (F) to form the "salt" lithium fluoride (LiF). By releasing one electron, lithium obtains the electron structure of the noble gas helium (He) and fluorine that of neon (Ne). When adding lithium fluoride to water, the salt dissolves to form lithium cations Li^+ and fluoride anions F^-.

The two borderline cases are distinguished by the following reaction equations in which a dot represents the potential binding electron of an element whereas a line between two atoms symbolizes a binding orbital filled with two binding electrons.

$A\cdot + \cdot A \rightarrow A - A$ (covalent bond)

and

$A\cdot + \cdot A \rightarrow A^+ A^-$ (ionic bond)

Normally, chemical bonds in molecules are not completely of ionic nature but are best described as covalent bonds with larger or smaller ionic contributions. The extent of ionic binding depends upon the differences in electronegativities of the binding partners.

1.1.2 Electronegativity of elements

The electronegativity describes the tendency of an element to attract the binding electrons in a chemical bond. In a molecule, the binding electrons on a time average are closer to the more electronegative reaction partner than the less electronegative (more electropositive) atoms. Thus, the chemical bonds obtain an ionic portion and the molecules have a permanent dipole moment.

According to a semi-empirical concept of *Pauling*[2], the electronegativities X of elements can be ascertained and the ionic fraction of a chemical bond A-B can be estimated. For that, the geometric mean of binding energies of the molecules A-A and B-B is calculated and subtracted from the binding energy determined experimentally for that molecule[3]. The idea behind this rationing is that the binding energies $D_{A\text{-}A}$ and $D_{B\text{-}B}$ of the chemical compounds A-A and B-B have only covalent binding energy, whereas $D_{A\text{-}B}$ is put together from covalent as well as ionic parts. Binding energies may be determined from electron spectra in the visible and UV section of the spectrum and are listed in reference books such as the "Handbook of Chemistry and Physics" (CRC Press). The difference Δ is described in Equation 1.1.

Equation 1.1 $$\Delta = D_{(A-B)} - [D_{(A-A)} \cdot D_{(B-B)}]^{1/2}$$

[2] Linus Carl Pauling, 1901 to 1994, Nobel Prize for chemistry in 1954 for his contributions to understand the nature of chemical bonds. Nobel Peace Prize in 1964 for his dedication against nuclear weapons.

[3] The binding energy is the energy needed to separate a chemical bond.

Pauling took the square root of Δ as a measure for the electronegativity difference between the atoms A and B.

Equation 1.2 $\sqrt{\Delta} = |X_B - X_A|$

In Equation 1.2, the index A denotes the element of the two with the lower electronegativity. It is customary to express Δ in electron volt (1 eV = 1.602 · 10^{-19} Joule). Pauling assigned an electronegativity value of $X_H = 2.2$ eV$^{1/2}$ to hydrogen[4].

Tablel.1 lists binding energies of a number of bonds between equivalent atoms. From these binding energies and tabulated values of binding energies between these atoms in molecules, values for $\sqrt{\Delta}$ can be calculated using Equations 1.1 and 1.2. This is shown for the reaction products of hydrogen with halogens in Table 1.2. In the case of H-F, for example, from $\sqrt{\Delta}$-values and $X_H = 2.2$ eV$^{1/2}$, the electronegativity of fluorine is computed to 3.96 eV$^{1/2}$.

The electronegativities of other elements are found accordingly. In theory, given that the binding energies are determined accurately enough, the electronegativities found are independent of the calculation path. In practice, however, published binding energies differ from source to source. This is also attributed to the fact that in molecules containing more than two atoms, the strength of chemical bonds will depend upon the molecules themselves. For example, the binding energy of the O-H bond in H-O-H (water) is 497 kJ/mole, whereas in CH_3-O-H (methanol) its value is only 440 kJ/mole. The electronegativities of the elements are listed in the periodic table in Figure 1.1.

1.1.3 Ionic contribution to a chemical bond

Molecules or sections of molecules with unsymmetrical distributions of electrons possess a dipole moment. The dipole moment is an aligned parameter, that is to say, it may be characterized as a vector. In the case of linear molecules, the dipole moment is positioned along the axis of the chemical bond. The dipole moment[5] $\vec{\mu}$ is the product between the electrical charges e and their separation distance $\vec{r}$.

Table 1.1: Binding energies of some molecules from identical atoms

Bond	Binding energy kJ/mol	eV
H-H	435	4.51
N-N	945	9.79
F-F	159	1.65
Cl-Cl	245	2.54
Br-Br	194	2.01
J-J	153	1.59
O-O	498	5.16
S-S	425	4.40
C-C	618	6.40

[4] Some sources state that the element fluorine was taken as a reference with an electronegativity of 3.98 eV$^{1/2}$.

[5] More on the determination of dipole moments is found in the course of this chapter.

Table 1.2: Calculation of electronegativities of halogen atoms

Compound	$\sqrt{D_A \cdot D_B}$	D_{AB} kJ/mol	Δ kJ/mol	Δ eV	$\sqrt{\Delta} = X_A - X_B$
H – F	263	570	307	3.18	1.8
H – Cl	326	431	105	1.09	1.0
H – Br	290	366	76	0.79	0.9
H – J	258	298	40	0.41	0.6

Equation 1.3 $\vec{\mu} = e \cdot \vec{r}$

In case of angulated molecules, the separate dipole moments of the chemical bonds add in a vectorial manner as shown in Figure 1.3.

A dipole consisting of a positive e^+ and a negative e^- elementary charge of 1.60219 10^{-19} C at a distance of one Ångstrom (= 10^{-10} m) has a dipole moment of 1.60219 10^{-29} $C \cdot m$. In anterior literature, dipole moments are reported in the unit Debye D (1 $C \cdot m = 2.988\ 10^{29}\ D$; 1 $D = 0.33467 \cdot 10^{-29}\ C \cdot m$).

In practice, the distribution of charges in a molecule is initially unknown. It is however possible to draw conclusions about the distribution of charges from measured dipole moments and also to estimate the share of ionic bonding in a chemical bond. For a molecule consisting of two atoms, this is done by multiplying the formal charges of the bond with the separation distance between the nuclei of the atoms in that chemical compound. In doing so, the dipole moment is calculated for the case that the bond is one hundred per cent ionic in nature. The measured dipole moment may be expressed as a percentage of the calculated dipole moment, thus revealing the ionic contribution.

Example: Hydrochloric acid, HCl:

Bond length: $1.275 \cdot 10^{-10}$ m. From Equation 1.3, a dipole moment of

$\mu = 1.275 \cdot 10^{-10}\ m \cdot 1.602 \cdot 10^{-19}\ C = 20.4 \cdot 10^{-30}\ Cm$ is calculated.

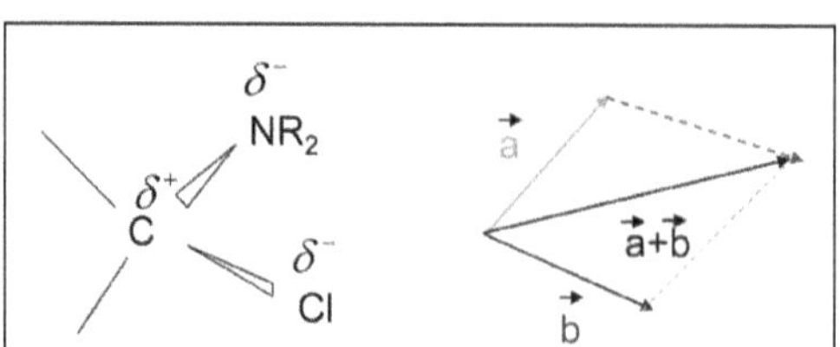

Figure 1.3: Vectorial addition of dipole moments along two different bonds in a molecule to a resulting dipole moment

The measured dipole moment is $3.44 \cdot 10^{-30}$ Cm. This corresponds to an ionic contribution to the bond of about 17 %.

Knowing the charge distribution in molecules is helpful when trying to understand or to foresee interactions between molecules[6]. In colloidal chemistry,

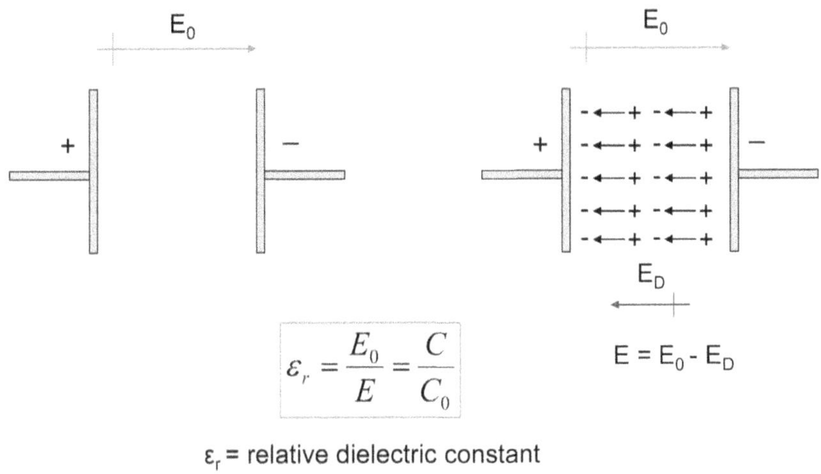

$$\varepsilon_r = \frac{E_0}{E} = \frac{C}{C_0}$$

Figure 1.4: Capacitor without (left) and with (right) a dielectric substance

this knowledge is beneficial for example when making assumptions about what section of a molecule will have the tendency to adsorb onto solid surfaces.

1.2 Physical interactions

Physical interactions between molecules arise either by virtue of their having a dipole moment (only molecules) or them being polarizable, i.e. that the electrons may be shifted relative to the atomic nuclei, thus making them dipolar (molecules and atoms). Substances that are either dipolar in nature or, otherwise, become dipolar when brought into an electric field, yet are not conductive, are called “dielectric”. The dielectric properties of substances are determined by measuring the capacitance of a capacitor with (C) and without (C_0) the substance.

1.2.1 Dielectric substances in a capacitor

In the simplest case, a capacitor (see Figure 1.4) is comprised by two metal plates that face each other without touching. The two plates can be oppositely charged by applying a voltage U. The capacitance of the capacitor tells how many charges Q appear on one of the two plates. The capacitance C is larger, the more charges are generated at a given voltage.

Equation 1.4 $$C = \frac{Q}{U}$$

[6] It is also helpful for chemists to anticipate possible chemical reactions that molecules might undergo.

An electric field with a field force (or field strength) E then fills the gap between the two plates. The electric field force is defined as the potential difference U between the plates separated by the distance d.

Equation 1.5 $$E = \frac{U}{d}$$

In the case that vacuum is in the gap between the plates, the electric field force E is directly proportional to the number of charges Q per unit area A of one capacitor plate. The proportionality constant ε_0 is called the “permittivity of vacuum”.

Equation 1.6 $$\frac{Q}{A} = \varepsilon_0 E$$

ε_0 has the value $8.854 \cdot 10^{-12}\ C \cdot V^{-1} \cdot m^{-1}$. When, instead of vacuum, a dielectric substance fills the gap between the capacitor plates, the number of charges rise by a factor which is characteristic for every individual substance. This is taken into account by introducing a relative dielectric constant ε_r as an expansion factor in Equation 1.6.

Equation 1.7 $$\frac{Q}{A} = \varepsilon_r \varepsilon_0 E \quad or \quad E = \frac{Q}{A\varepsilon_r \varepsilon_0}$$

By substituting the potential U in Equation 1.4 by $E \cdot d$ (Equation 1.5) and further substituting E by $Q/A\ \varepsilon_r\ \varepsilon_0$ (Equation 1.7), a new expression for the capacitance of a capacitor filled with a dielectric is found.

Equation 1.8 $$C = \varepsilon_r \varepsilon_0 \frac{A}{d}$$

Since, by definition, ε_r in vacuum is unity, ε_r may consequently be determined from the ratio of the capacitance of a capacitor with (C) and without (C_0) the presence of the dielectric. “Without a dielectric” means either in vacuum or, adequately enough, in air, since air has a low density and since the molecules in the atmosphere are barely polarizable.

Equation 1.9 $$\varepsilon_r = \frac{C}{C_0} = \frac{E_0}{E}$$

According to Equation 1.8, the introduction of a dielectric always leads to an increase in capacitance or, respectively, to a drop of the electric field force in the capacitor (Equation 1.7).

The reason for this lies in the “polarization” P of the dielectric. Polarization depicts the formation of an electric field within the capacitor opposite to the applied outer electric field, thereby weakening it. The total effect is composed of a contribution coming from the displacement of the electrons relative to the nuclei of the dielectrics (induced displacement or electron polarization P_i) and a second component stemming from the orientation of dipolar molecules (orientation polarization P_0).

Equation 1.10 $$P = P_i + P_0$$

Both types of polarization are schematically presented in Figure 1.5. When the polarization is standardized to one mole of substance, it is called "molar polarization" P_M.

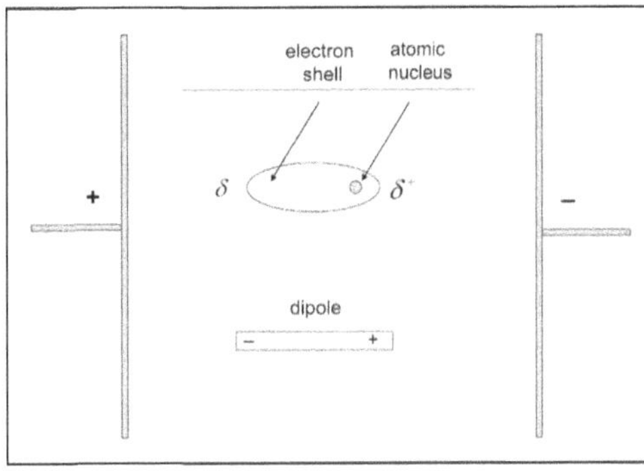

Figure 1.5: Emergence of electron polarization in a capacitor by displacement of the electrons relative to the nuclei (top) and orientation polarization by the orientation of dipolar molecules in the electric field of the capacitor (bottom).

The electron polarization happens very rapidly and is independent of the temperature since it relies only upon the mobility of the electrons in a molecule or atom. Electrons follow any changes of the electric field instantaneously because of their small mass. Typically, electrons are excited by electromagnetic waves with a frequency of 10^{14} Hertz. This corresponds to the region of visible light (approximately $4 - 7 \cdot 10^{14}$ Hz).

This is different in the case of orientation polarization. Orientation happens more slowly and is temperature dependent. As opposed to electrons, molecules are heavy so that they possess inertia, letting them follow changes in the electric field only with a certain delay time. So, when an alternating electric field is applied to a capacitor, it will depend upon the frequency of the field change if dipolar molecules have enough time to adjust to the field or not. Low molecular weight materials such as common solvents need about 10^{-12} seconds time to follow the field. That's why electric field frequencies of 10^5 Hz are usually applied when the contribution of orientation polarization to the relative dielectric constant is to be accessed. The temperature dependency of the orientation polarization is due to the thermal fluctuation of molecules, which is known as the "Brownian motion". The thermal movement counteracts the orientation of the molecules.

Both the electron polarization and the orientation polarization create an electric field with a magnitude of $P/(3 \cdot \varepsilon_0)$ that opposes the applied electric field, whereby P has the physical dimension of a charge density ($C \cdot m^{-2}$) or a volume based dipole moment ($C \cdot m \cdot m^{-3}$), respectively.

1.2.2 Electron polarization

The electron polarization is described by the "Clausius-Mosotti" equation (Equation 1.16). The starting point for its derivation [1] is to consider the dielectric substance as a single dipole consisting of many dipoles, each having an induced dipole moment of μ_i.

Equation 1.11 $$P_i = \frac{N}{V}\mu_i = \frac{L\rho}{M}\mu_i$$

N = number of molecules within the volume V with a dipole moment of μ_i
L = Avogadro's number
M = molecular weight
ρ = molar volume

The induced dipole moment is proportional to the effective field strength F. The proportionality constant is the polarizability α.

Equation 1.12 $$\mu_i = \alpha \cdot F$$

α has the dimension of a reciprocal volume. The effective field strength within the dielectric substance is the vector sum of the externally applied electric field with the field strength E and the opposed field.

Equation 1.13 $$F = E + \frac{P}{3\varepsilon_0}$$

From Equations 1.12 and 1.13 follows for the induced dipole moment

Equation 1.14 $$\mu_i = \alpha\,(E + \frac{P}{3\varepsilon_0})$$

The polarization is expressed by

Equation 1.15 $$P = (\varepsilon_r - 1) \cdot \varepsilon_0 \cdot E$$

Solving Equation 1.11 for μ_i and Equation 1.15 for E and inserting these expressions for μ_i and E into Equation 1.14, followed by rearrangement leads to the Clausius-Mosotti equation [7].

Equation 1.16 $$\frac{\varepsilon_r - 1}{\varepsilon_r + 2}\frac{M}{\rho} = \frac{L\alpha}{3\varepsilon_0}$$

The Clausius-Mosotti equation enables the determination of the polarizabilities of dielectric substances by measuring their relative dielectric constants. As stated

[7] The Clausius-Mosotti equation is also known under the names Lorentz-Lorenz equation or Maxwell equation.

in the introduction to this chapter, dispersive interactions between molecules can be calculated from their polarizabilities.

Alternatively to capacity measurements in a capacitor, the portion of the dielectric constant coming from electron polarization can be calculated from the refractive indexes n of dielectric substances. According to Maxwell's theory of electromagnetic radiation:

Equation 1.17 $$\varepsilon_r = n^2$$

1.2.3 Orientation and molar polarization

The Clausius-Mosotti equation regards only the contribution of the displacement of the electrons to the molar polarization P_M. Since the total dipole moment may also have a contribution resulting from the orientation polarization, this part must be added to the induced polarization of the electrons. As already stated, the orientation polarization decreases with increasing temperature. Using the Boltzmann equation and applying a number of simplifying assumptions, an expression for the molar orientation polarization is found [1].

Equation 1.18 $$\frac{\varepsilon_r - 1}{\varepsilon_r + 2}\frac{M}{\rho} = \frac{L}{3\varepsilon_0}\frac{\mu^2}{3kT}$$

Equation 1.10 shows that the total polarization is put together by parts coming from induction and from orientation. Hence, Equation 1.16 and 1.18 are combined to the so called Debye equation:

Equation 1.19 $$\frac{\varepsilon_r - 1}{\varepsilon_r + 2}\frac{M}{\rho} = P_M = \frac{L}{3\varepsilon_0}(\alpha + \frac{\mu^2}{3kT})$$

Two methods are used to ascertain polarizabilities α and dipole moments μ. Either the relative dielectric constants are measured at different temperatures and P_M is plotted against the reciprocal temperature $1/T$. Then, a straight line is found with the slope of $\mu^2/3k$ and an intercept with the y-axis at $L\alpha/3\,\varepsilon_0$. The dipole moment can then be calculated from the slope, whereas the polarizability is found from the intercept. Alternatively, ε_r may be calculated according to Equation 1.17 from the refractive index and then the polarizability α can be found from Equation 1.16. Subsequently, the dipole moment can be calculated with Equation 1.19 from the dielectric constant ε_r measured with the capacitor and from the α-value acquired via the refractive index.

Table 1.3 gives an overview of the dielectric constants of a number of solvents widely used in paint formulations. The dielectric constants of liquids are also a good measure for their ability to stabilize ions by solvation. The higher they are, the greater the ability of the liquids is to dissolve salts. The relative dielectric constant of water, for example, lies at about 80.

Table 1.3: Relative dielectric constants of some organic solvents

Compound	rel. DK	Compound	rel. DK
Cyloaliphatic carbohydrates		**Ethers und glykole ethers**	
Dekaline	2.20	Ethylglykole	13.70
Cyclohexane	2.00	Butyldiglykole	11.00
Aromatic carbohydrates		Butylglykole	9.20
ortho-Xylene	2.57	Tetrahydrofurane	7.60
meta-Xylene	2.37	**Esters**	
para-Xylene	2.30	Butyldiglykol acetate	7.00
Ethylbenzene	2.30	Ethylacetate	5.60
Alcohols		n-Butylacetat	4.50
Isopropanol	26.00	**Ketones**	
Ethanol	22.40	Cyclohexanon	18.30
n-Propanol	22.20	Methyl-n-propylketone	15.40
Isobutanol	18.40	Methyl isobutyl ketone	13.11
n-Butanol	18.20	**Glykoles**	
Cyclohexanol	15.00	Ethylenglykole	41.20
Tert.-Butanol	10.90	Diethylenglykole	32.00
Isooktylalcohol	10.00	Triethylenglykole	24.00

1.3 Energies and forces of attraction

In case the polarizabilities and the dipole moments of dielectric substances are known, the energies of attraction between congeneric or between disparate molecules can be calculated using rather simple equations.

The following interactions are distinguished

- Interactions between two dipoles (dipole-dipole or "Keesom" interaction)
- Interactions between a dipole and a polarizable substance (induced dipole or "Debye" interaction)
- Interactions between two polarizable substances (dispersive or "London-van der Waals" interaction)

All of these interactions are extremely distance-dependent. In fact, the energies of interaction are proportional to the reciprocal value of the distance between the molecules (or atoms) to the power of six. In general terms, energy is the ability to perform work. Energy and work are two expressions for one and the same thing. When two molecules (or atoms) are a distance of r to each other and have an interaction energy E_r, then E_r is the work that must be applied in order to separate them to a distance where they no longer attract each other. Theoretically, this is the case at an infinite distance. In practice, taking the high distance dependency into account, interactions decay to small values at only moderate separations already. Since E is proportional to $1/r^6$, by doubling the separation distance of the two molecules, the interaction energies diminish to 1.5 % of the original value. Attractive interactions get a negative sign, whereas repulsive interactions are depicted by their positive sign.

An even more descriptive quantity than "energy" is "force". Energy is defined as the product between force and distance. Therefore, by mathematical differentiation of energy with respect to the distance, the attractive (or repulsive) force is obtained.

1.3.1 Dipole-dipole interaction

The dipole-dipole (Keesom) interactions come into place when two permanent dipoles align in such a way that the negatively polarized part of the first molecule shows towards the positively polarized part of the second molecule, so that an attractive force is generated along their dipole axes. Dipole-dipole interactions are only possible between molecules and not between atoms since the latter have no permanent dipole moment.

Equation 1.20
$$E_{DD} = -\frac{1}{(4\pi\varepsilon_0)^2}\,\frac{1}{r^6}\,\frac{2\cdot\mu_A^2\cdot\mu_B^2}{3kT}$$

The dipole-dipole interaction energy is, according to Equation 1.20, proportional to the square of the dipole moments of both molecules A and B. When dipole-dipole interaction takes place between alike molecules, the interaction energy is proportional to the power of four of the dipole moment, thus proportional to μ^4. The inverse dependency towards the separation distance of the molecules to the power of six was already mentioned. The factor $1/(4\pi\,\varepsilon_0)^2$ arises in the derivation of Equation 1.20 due to geometric considerations of the interactions of point charges in a three-dimensional space. The factor $1/kT$ results from considering the thermal movement of the molecules which counteracts their alignment.

1.3.2 Induced dipole interactions

The induced dipole (Debye) interaction comes into effect when a dipolar molecule changes the electron density distribution of a neighbouring molecule or atom relative to its nucleus. Debye interactions therefore take place either between molecules or between molecules and atoms, yet never between atoms alone. The interaction energies in this case are proportional to the sums of the products of the polarizabilities and the square of the dipole moments of the molecules A and B.

Equation 1.21
$$E_{ind} = -\frac{1}{(4\pi\varepsilon_0)^2}\frac{1}{r^6}(\mu^2_A \cdot \alpha_B + \mu^2_B \cdot \alpha_A)$$

In the specific case that the interactions take place between equal molecules, the term in brackets of Equation 1.21 is substituted by $2 \cdot \mu^2\alpha$. The remarks to the other factors standing in front of the brackets that were made in the preceding section apply here also.

1.3.3 London-van der Waals interaction

The dispersive (London-van der Waals) interaction originates when fluctuations of the electrons in one molecule or atom couple with the electron fluctuations in another molecule or atom. When the electron density distributions oscillate in phase with a frequency ν_0, then a short termed attraction amongst the molecules or atoms is generated. The energy of this attraction is proportional to the product of the polarizabilities of the molecules or atoms involved. The proportionality is given by Planck's constant h $(h = 6.626 \cdot 10^{-34}\ J \cdot s)$

Equation 1.22
$$E_L = \frac{1}{(4\pi\varepsilon_0)^2}\frac{1}{r^6}\frac{3}{4}\,h\nu_0\alpha_A\alpha_B$$

In the case of identical molecules or atoms, $\alpha_A\alpha_B$ is replaced by α^2.

The characteristic frequency ν_0 may be determined with the help of Cauchy's dispersion formula by measuring the refractive index n of the chemical at different light wave frequencies ν:

Equation 1.23
$$\frac{n^2+2}{n^2-1} = Const. \cdot \nu_0^2 - Const. \cdot \nu^2$$

Equation 1.23 represents a straight line with a slope of *Const.* and an intercept with the ordinate at $Const. \cdot \nu_0^2$. By plotting the left hand side of the equation $(n^2 + 2)/(n^2 - 1)$ against the square of the frequency of the light, the characteristic frequency ν_0 may be assessed. Since the frequency dependency of the refractive index or any other physical property of matter is termed "dispersion", the Londen-van der Waals interaction is also called "dispersive interaction".

If dispersive interactions between different atoms or molecules were to be calculated, one must theoretically consider that they might have different characteristic frequencies. Some authors propose to use 1 MJ/mole as a general value for $h\,\nu_0$. In that case, the characteristic frequency would be assumed to lie in the ultraviolet region of the spectrum at a frequency of $2.5 \cdot 10^{15}$ Hertz, corresponding to a fixed wavelength of 120 nm. Provided the characteristic frequencies are known, calculations could be carried out by using a mean value of the characteristic frequencies. A third possibility is expressing the term $h\,\nu_0$ by the ionization potentials I_A and I_B of the molecule kinds A and B:

Equation 1.24
$$E_L = \frac{1}{(4\pi\varepsilon_0)^2}\,\frac{1}{r^6}\,\frac{3I_AI_B}{2(I_A + I_B)}\,\alpha_A \cdot \alpha_B$$

Ionization energies are the energies necessary to remove an electron from the most outer orbital of a molecule or an atom and bring it to an infinite distance. For electrons in organic substances, they lie in the order of magnitude of about $6 \cdot 10^5$ J/mole, or 10^{-18} J for a single electron, respectively.

1.3.4 Born interaction

Up to now, only attractive forces between atoms or molecules have been considered. There are, however, also repulsive forces. If molecules could approach each other unhindered, their electron shells would overlap so that the orbitals would be populated with more than two electrons. In that way, there would be electrons with identical quantum numbers in some of the orbitals. This contradicts the "Pauli exclusion principle" which explains why one substance cannot penetrate another. One way to account for this is to assume that, at a certain minimal distance, an infinitely large repulsive energy comes into place. According to Born, the repulsive energy E_B may also be expressed as a high power function of the distance r.

Equation 1.25
$$E_B = \frac{B}{r^m}$$

In Equation 1.25, m is in the order of magnitude between 9 and 12. A value of 12 is commonly used for m, whereas B is a constant.

1.3.5 Total interaction energy

The total interaction energy E_{tot} at a certain separation distance between molecules or atoms is given by the sum of all attractive and repulsive energies at that distance. Equations 1.20 to 1.22, 1.24 and 1.25 therefore yield:

Equation 1.26
$$E_{tot} = E_{DD} + E_{ind} + E_L + E_B$$

Accordingly, by adopting the complete expressions of the individual interaction energy contributions:

Equation 1.27
$$E_{tot} = -\frac{1}{(4\pi\varepsilon_0)^2 r^6}\left[\frac{\mu^2_A\mu^2_B}{3kT} + \mu^2_A \cdot \alpha_B + \mu^2_B \cdot \alpha_A + \frac{3}{4}\,h\nu_0\alpha_A\alpha_B\right] + \frac{B}{r^{12}}$$

As mentioned earlier, "energy" is the product of "force" times "distance". The resulting attractive force F_{tot} is therefore found by differentiating Equation 1.27 with respect to the distance r.

Equation 1.28

$$F_{tot} = \frac{dE_{WW}}{dr} = \frac{6}{(4\pi\varepsilon_0)^2 r^7} \left[\frac{\mu_A^2 \mu_B^2}{3kT} + \mu_A^2 \cdot \alpha_B + \mu_B^2 \cdot \alpha_A + \frac{3}{4} h\nu_0 \alpha_A \alpha_B \right] - 12 \frac{B}{r^{13}}$$

Equation 1.28 in comparison to Equation 1.27 illustrates that attractive forces between molecules and/or atoms are even more distance dependent than the corresponding energies. Due to the mathematical operation of differentiation, the attractive forces obtain a positive sign, which accounts for the fact that negative forces do not exist[8].

1.3.6 Lennard-Jones potential

Unfortunately, Equation 1.27 is however not applicable without further ado, since B is unknown. Lennard-Jones [2] expressed Equation 1.27 in the general form:

Equation 1.29 $$E_{tot} = -\frac{A}{r^6} + \frac{B}{r^{12}}$$

whereby A and B both depend upon σ, the distance of the centres of species A and B at which the attractive energy is zero, so that they neither attract nor repel each other, and ε, the highest attractive energy.

Equation 1.30 $$A = 4 \cdot \varepsilon \cdot \sigma^6$$

Equation 1.31 $$B = 4 \cdot \varepsilon \cdot \sigma^{12}$$

By substituting Equation 1.30 and 1.31 into 1.29 and differentiating with respect to the distance r and then fixing E_{tot} to zero, the distance r_m is found, at which the highest attractive energy is in place.

Equation 1.32 $$r_m = \sqrt[6]{2} \cdot \sigma$$

Figure 1.6 shows the principle dependency of the total interaction energy as a function of the distance between two atoms or molecules. The figure therefore also represents the total interaction energy between atoms or molecules according to Equation 1.27. At large separation distances, the atoms or molecules undergo little or no interaction. However, when approaching each other, the attractive interaction energy obtains progressively larger, negative values. At a distance r_m, the interaction energy runs through a minimum, so that the attractive energy is highest at that point. As the distance becomes even shorter, the attraction becomes less and finally turns into a repulsive interaction at further approach. The con-

[8]....at least not in technology.

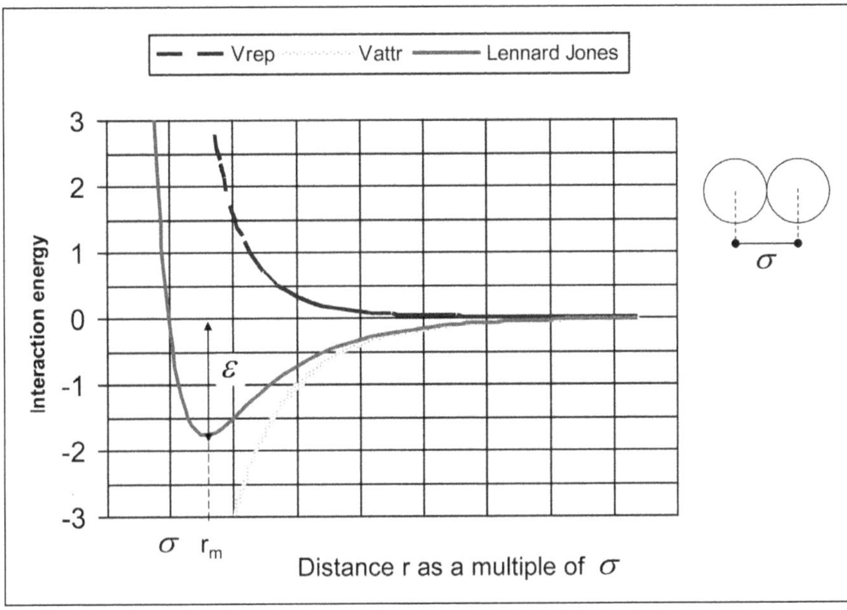

Figure 1.6: Lennard-Jones potential between two atoms

sequence is that the two atoms or molecules that are associated vibrate around a mean separation distance r_m.

The Lennard-Jones potential was used to describe the distance dependent interaction between molecules of gas. In the case of the noble gas argon, for example, $\sigma = 3.405$ Ångstrom was found to be the distance at which the atoms neither attract, nor repel each other and the maximum attraction energy of $\varepsilon = 165.3 \cdot 10^{-21}$ Joules was calculated for two argon atoms at a separation distance of $r_m = 3.822$ Ångstroms [3].

In the further course of this book it will be shown that colloidal particles exhibit similar interaction energy curves like atoms and molecules as long as both attractive and repulsive forces are in place.

1.4 Hydrogen bonds

A further physical interaction type that, however, may only occur between certain molecules, is the hydrogen bond. Hydrogen bonds are characterized by one molecule acting as a proton donor and another as a proton acceptor. Within the proton donor molecule, a hydrogen atom must form a chemical bond with an atom that has a high electronegativity. The proton acceptor, on the other hand, should have

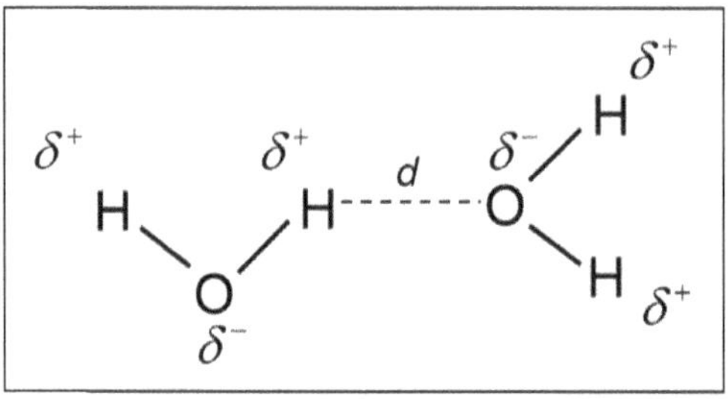

Figure 1.7: Polarization of electrons in a hydrogen bond

lone electron pairs. As an example, Figure 1.7 shows how two water molecules are polarized while forming a hydrogen bond. The molecules arrange in a way that the overlap between the protons of the proton donor and the free electron pair of the proton acceptor maximizes. In liquid water, clusters of approximately nine individual molecules exist. The high conductivity of water is, for example, attributed to the switching of hydrogen bonds to chemical bonds, whereby positive charges are transferred without mass transport being involved.

Strong hydrogen bonds of the type *X-H----Y* occur mainly when the atom *X* is either an oxygen, a nitrogen or a halogen atom (fluorine, chlorine, bromine or iodine) whereas the atom *Y* may be oxygen, nitrogen, sulfur or a halogen atom. Hydrogen bonds play an important role in many processes taking place in the animated and non-animated nature. In colloidal chemistry, hydrogen bonds are often determining factors as well because of their strength and their special alignment.

There is as yet no coherent method for calculating interaction energies from hydrogen bonds between molecules. They usually lie in the order of about 10 to 50 KJ/mole. The *F-H------F* bond is the most powerful hydrogen bond. A bonding energy of 160 to 170 KJ/mole is attributed to it.

1.5 Range of physical interaction energies

Polarizabilities, dipole moments and values of $h \cdot \nu_0$ for a number of molecules as well as the contributions of attraction energies coming from the different mechanisms are presented in Table 1.4. It shows that dispersive forces always play a major role in intermolecular interactions. In contrast, appreciable dipole-dipole interactions only come into being when dipole moments exceed 1.3 Debye ($= 4.3 \cdot 10\text{-}30\ C \cdot m$). Induced dipole interactions are very weak in all cases.

The noble gases naturally have only dispersive interactions, whereas, on the other hand, water has a high contribution of dipolar interactions. Next to this, water molecules are attracted to each other by hydrogen bonds.

An indication for the interaction between molecules or atoms is their boiling points. Table 1.5 lists polarizabilities, dipole moments and boiling points of a few selected chemical compounds. Neither of the noble gases (He, Ne, Ar, Kr, Xe) have a permanent dipole moment. However, their polarizabilities increase

Table 1.4: Contributions of dipole-dipole, induced dipole and dispersive interactions to the attractive energies of some small molecules [4]

Compound	Dipole moment μ (10^{-30} Cm)	Polarisability α (10^{-30} m^3)	Energy $h\nu_0$ (eV)	Orientation $^2/_3\ \mu^4/kT$	Induction $2\mu^2\alpha$	Dispersion $^3/_4\ \alpha^2 h\nu_0$
CO	0.4	1.99	14.3	0.0034	0.057	67.5
HJ	1.27	5.40	12.0	0.35	1.68	382
HBr	2.61	3.58	13.3	6.2	4.05	176
HCl	3.45	2.63	13.7	18.6	5.4	105
NH_3	5.02	2.21	16.0	84	10	93
H_2O	6.16	1.48	18.0	190	10	47
He	0	0.20	24.5	0	0	1.2
Ar	0	1.63	15.4	0	0	52
Xe	0	4.00	11.5	0	0	217

with their molar mass. Their boiling points rise extremely as a function of their weight. Methane, which also has no permanent dipole moment, is, with respect to its polarizability of $25.9 \cdot 10^{-31}\ m^3$, comparable to the noble gas krypton with a polarizability of $24.6 \cdot 10^{-31}\ m^3$. Consequently, their boiling points of 111.7 K and 119.9 K are also very similar. In the row isobutane, isobutylene to trimethylamine, the dipole moments increase from $0.44 \cdot 10^{-30}\ Cm$ to $2.23 \cdot 10^{-30}\ Cm$ at a comparable polarizability level. The boiling points, however, lie between 263 K and 278 K and seem to be almost independent of the dipole moments. This leads to the conclusion that dipolar interactions have only little influence on the overall interaction between molecules. More so, the influence of dipolar interactions becomes less, the larger the molecules are.

Deviations in the boiling behaviour of molecules are, however, evident with molecules capable of forming hydrogen bonds. In Table 1.5, these compounds are ammonia, water and methanol. Ammonia and krypton have about the same polarizabilities. In light of the afore mentioned relations, the dipole moment of ammonia does not suffice to explain the differences in the boiling points of ammonia (240 K) and krypton (119.9 K). The same is true for the comparison between water (373 K) and argon (87.3 K). The influence of hydrogen bonds also becomes evident when comparing the boiling points of methyl fluoride (195 K) and methanol (338 K). In this case, the ability to form hydrogen bonds leads to a boiling temperature increase of 143 K at comparable polarizabilites and dipole moments.

Table 1.5: Polarizabilities, dipole moments and boiling points of some atoms and molecules

Compound	Chemical formula	Molecular weight [g/mol]	Polarisability α [m³] · 10^{31}	Dipole moment μ [Cm] · 10^{30}	Boiling pt. K
Helium	He	2	2.03	0.00	4.2
Neon	Ne	10	3.92	0.00	27.3
Argon	Ar	18	16,30	0.00	87.3
Krypton	Kr	36	24.60	0.00	119.9
Xenon	Xe	54	40.10	0.00	165.1
Methane	CH_4	16	25.90	0.00	111.7
Ammonia	NH_3	15	23.40	4.90	240.0
Water	H_2O	18	15.30	6.20	373.0
Methylfluoride	CH_3F	34	38.40	6.04	195.0
Methanole	CH_3OH	32	29.70	5.67	338.0
Isobutane	$(CH_3)_3CH$	58	83.60	0.44	263.0
Isobutylene	$(CH_3)_2 = CH_2$	56	83.60	1.63	267.0
Triethylamine	$(CH_3)_3N$	59	80.90	2.23	278.0
Chlorbenzene	C_6H_5Cl	112	53.50	5.42	248.9
n-Hexane	C_6H_{14}	86	119.00	0.00	341.7

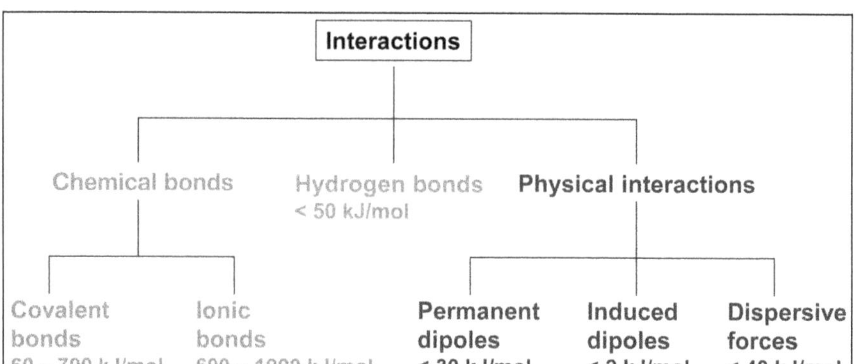

Figure 1.8: Comparison of typical energy contents of one mole (6.0231 x 10^{23}) of physical and chemical bonds , respectively

Figure 1.8 shows typical ranges of energies associated with the different types of chemical and physical bonds. As a rule of thumb, chemical bonds are about ten times stronger than physical bonds. In case of physical bonds, the binding energies decline from dispersive over dipole-dipole to induced dipole interactions. Hydrogen bonding energies normally lie at the level of dispersive interactions, although with a potential towards markedly higher values.

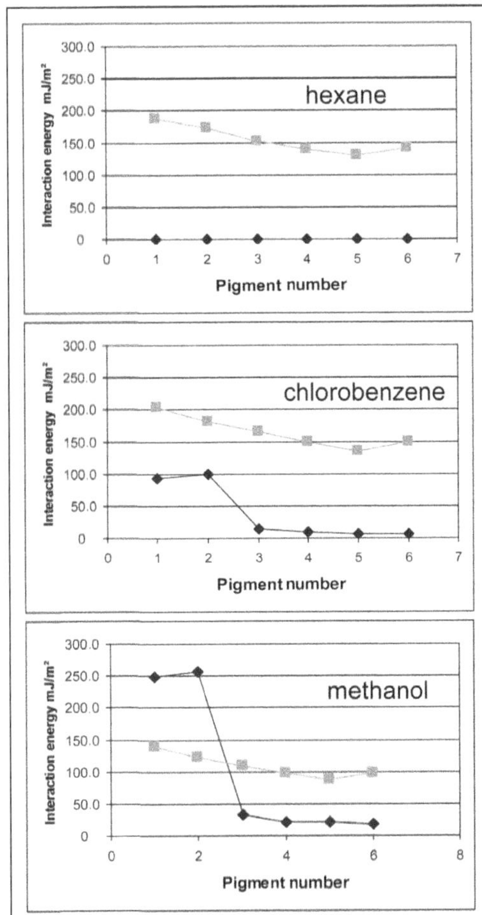

Figure 1.9: Contributions of dispersive (rectangles) and polar (rhombuses) interactions to the wetting of the pigments 1: TiO_2, 2: γ-Fe_2O_3, 3: γ-quinacridone, 4: β-copper phthalocyanine, (isometric), 5: chlorinated copper phthalocyanaine (Cl16CuPc), 6: β-copper phthalocyanine (rod shaped) with
a) n-hexane
b) chlorobenzene
c) methanol
re-drawn from [5]

1.6 Interactions at interfaces

At the interface between solid and liquid phases, the interactions described above also come into action. *Schroeder* [4] estimated dipole moments and polarizabilities of inorganic and organic pigment surface molecules in the liquids n-hexane, chlorobenzene and methanol and calculated the contributions of the different interactions from these values (see Figure 1.9). The pigments employed were titanium dioxide, γ-iron oxide red, γ-quinacridone, β-copper phthalocyanine (isometric), polychloro copper phthalocyanine ($Cl_{16}CuPc$) and a rod shaped β-copper phthalocyanine. The dipole moments and polarizabilities of n-hexane and chloro-benzene are also listed at the end of Table 1.5.

Schroeder found (viz. Figure 1.9) that all pigments he examined only underwent dispersive interactions in the energy range between 125 and 250 mJ/m² when immersed into n-hexane. The dispersive interactions with inorganic pigments were more pronounced than in the case of organic pigments. With chlorobenzene, the dispersive interactions of all the pigments were more or less identical to those in n-hexane. Yet, only the two inorganic pigments exhibited additional polar interactions of approximately 100 mJ/m². The organic pigments underwent only slight polar interactions with chlorobenzene. In methanol, the dispersive interactions reduced to values between 90 (organic pigments) and140 mJ/m² (inorganic pigments). Whereas inorganic pigments possessed additional polar interactions of about 250 mJ/², the equivalent values for organic pigments merely laid between 20 and 30 mJ/m². This exemplifies that dispersive interactions contribute largely to interactions at the pigment-liquid interface. Furthermore, it shows that dipole-dipole interactions are limited to polar pigment surfaces, naturally, only provided that the liquids themselves (viz. Equation 1.20) have a high dipole moment or, as in the case of methanol, are able to form hydrogen bonds. Hydrogen bonds can only come into place if the pigments possess hydroxyl groups or other groups, such as –S-H or =N-H, that are capable of forming hydrogen bonds. These can come from the pigments themselves or from inorganic surface treatments.

1.7 Literature

[1] G. Wedler, Lehrbuch der Physikalischen Chemie, Wiley-VCH, 5. Auflage, Weinheim 2004

[2] J. Lennard-Jones, Proc. Royal Soc. 106, (1924) 463

[3] A. Michels, H. Wijker, H. K. Wijker, Physica 15 (1949) 627

[4] J. Schröder, Colloid Interface Sci. 72, No. 2 (1979) 279

2 Properties of pigments and fillers

When dispersing, pigments and fillers, agglomerates are broken down and turned into primary particles. For that to happen, the attractive forces between the particles have to be overcome. Pigment agglomerates are held together by London-van der Waals interactions. Inorganic and organic surface treatments modify the attractive forces so that the assembly of agglomerates is affected by them. The relationship between the Hamaker constant of an organic surface treatment, agglomerate structure and the dispersibility of pigment agglomerates is explained in this chapter. The following important propositions prevail:

1. Within this book, no difference is made between pigments and fillers (also called extenders in Europe). According to DIN EN ISO 4618, pigments and fillers are distinguished only in that pigments are used for "optical, protective or decorative" properties, whereas fillers have "a refractive index usually less than 1.7" and are utilized because of their "physical or chemical properties". Otherwise, both materials consist of particles which are insoluble in the media they are used in. Solely that matters from a colloid chemistry point of view.
2. A strict disparity must be made between dispersion processes, in which merely physical interactions between the particles must be overcome and milling processes. In the latter case, chemical bonds are broken, thus making much higher power inputs necessary when compared to dispersions. At least a tenfold amount of energy is needed to disrupt chemical rather than physical bonds.

2.1 Dispersing and milling

In the production of polymeric composite materials such as paints, plastics, rubbers and synthetic fibres, pigments and fillers are nearly always employed so that dispersing becomes essential. Pigments and fillers normally consist of (sub-) microscopic particles with mean particle diameters ("particle sizes") between a few nanometres up to some micrometres. During production they normally evolve as dry powders. In a colloid chemical sense, they are actually not different from one another. Whereas pigments are used to add colour to polymeric materials or to achieve specific properties such as corrosion protection, fillers are incorporated with the aim to enhance applicational or mechanical composite material properties, or, otherwise, simply to reduce their price. Given that the pricing is volume

Jochen Winkler: Dispersing Pigments and Fillers

ISBN 978-3-86630-852-7

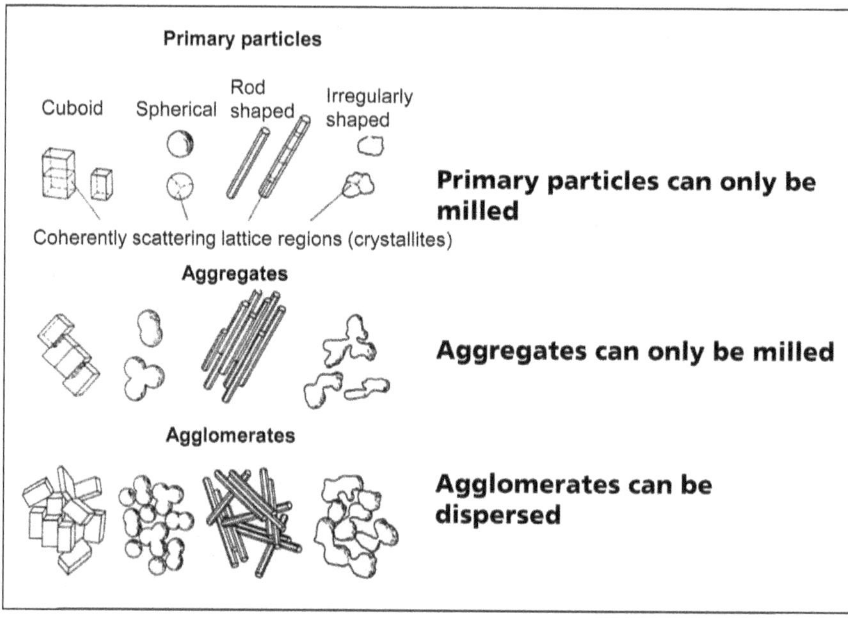

Figure 2.1: Pigment model according to DIN 53 206

based, the latter is of course only possible if the volume based price of the filler is lower than the cost for the equivalent volume of solid polymer.

According to the standard DIN 53 206, primary particles, agglomerates and aggregates should be distinguished from one another (Figure 2.1).

Primary particles are the smallest constituents in pigment or filler powders. They may consist of a number of crystallites, for example in case the pigments are made by calcining a precipitated precursor. An example for this procedure is titanium dioxide pigments. In the so called "sulfate route", titanium dioxide (TiO_2) pigments are produced by calcining a precipitated hydrous oxide of titanium ($TiO(OH)_2$) at temperatures of around 850 °C. If the onset of crystallization occurs simultaneously at different parts of the hydrous oxide, then the primary particles obtained have sections in which the crystal lattices are orientated in different directions. These domains are called crystallites. During calcination, primary particles that are formed may also sinter (bake) together to form larger particles. These are called **aggregates**. On the contrary, particles that are not connected by chemical bonds but merely by physical bonds instead are called **agglomerates**. Agglomerates have a surface area which is as large as the sum of the areas of the particles of which they are put together. Aggregates, on the other hand, have a lower surface area because the primary particles are conjoined in a platy way. Apart from calcination steps, which are often important in the production of inorganic pigments, aggregates can also derive from crystallization steps, which are predominantly typical for the production of organic pigments.

When dispersing, agglomerates are broken down into primary particles as well as smaller agglomerates. In doing so, only physical bonds are ruptured. It was shown in Chapter 1.5 that these are in the energy range between 40 and 50 KJ/mole. This means that one mole (= $6.02 \cdot 10^{23}$) of physical bonds has an energy content of 40,000 to 50,000 joules or, respectively, that this amount of energy is necessary to break this number of physical bonds. If aggregates are milled, i.e. if chemical bonds are broken, then an energy input of approximately 500 to 1000 KJ/mole is required, which is approximately tenfold. This case is termed **milling** or **grinding**.

2.2 Particle size determination

As Figure 2.1 suggests, pigment particles tend not to be of spherical shape, but rather possess a "form factor". This means that they are mostly irregularly shaped or may be rod shaped or platy. Nevertheless, it is customary to assign "mean particle sizes" to the different pigments. Often, the expression "equivalent particle diameter" is used, meaning the diameter of a spherical particle having the same volume or identical properties (such as sedimentation velocity) as the typical pigment particle. Of course, the primary particles of pigment powders are not of uniform size but rather exhibit a range of sizes. Pigment particle size distributions may be accessed in a number of different ways. The formerly common method of **sedimentation rate analysis** in a liquid, either under the influence of gravity or in a centrifugal field has been largely replaced by **laser light scattering** and **laser light diffraction** techniques. Laser light scattering principles are especially useful for very small particles with mean particle sizes of less than 200 nm, whereas laser diffraction techniques give more meaningful results in the case of larger particles. The advantage of both methods lay in the swiftness of the measurements. The disadvantage is, however, that the measurements are less accurate, especially for widely spread particle diameter distributions or when the mean particle sizes lay outside of the optimum range of the measuring principle employed. Alternative methods are based upon the determination of the inertia of the particles (electro-acoustic methods), the displacement of electrolyte solutions in electric fields ("electronic zone sensing" with the "Coulter Counter"), or the calculation of mean particle sizes from specific surface areas of the powders. Specific surface areas of pigments and fillers may be accessed by nitrogen gas adsorption using the one-point method according to Haul and Duembgen [1] (sometimes called the "BET-surface area" after Brunauer, Emmet and Teller). Assuming the particles to be spherical and non-porous, the mean particle diameter d may easily be estimated from the specific surface area A_{spec} of the particles and from their density ρ.

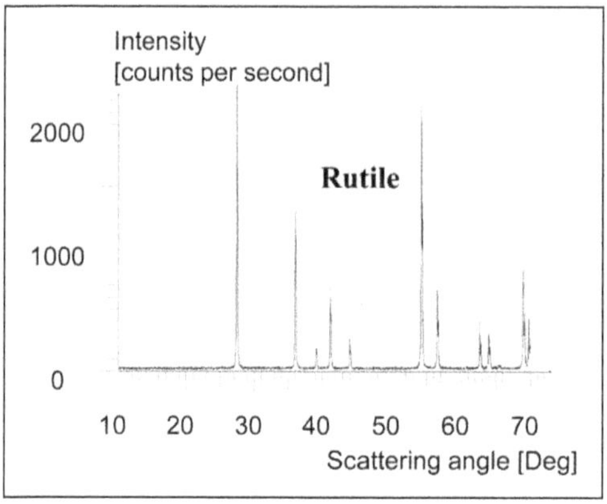

Figure 2.2: X-ray diffraction spectrum of a titanium dioxide pigment in the rutile modification

Equation 2.1 $$d = \frac{6}{A_{spec} \cdot \rho}$$

In the case of nano-sized pigments that, due to their fineness, consist of only one single crystallite, the measurement of crystallite sizes with the aid of **X-ray diffraction** is convenient. The mean crystallite size τ may be acquired by the Scherrer equation (Equation 2.2) from the X-ray wave length λ, the scattering angle Θ and the "half width" β of the X-ray peak.

Equation 2.2 $$\tau = \frac{0{,}89 \cdot \lambda}{\beta \cdot cos\theta}$$

Figure 2.2 shows the X-ray diffraction spectrum of a titanium dioxide pigment in the rutile modification. The Scherrer analysis is normally performed automatically by the software of X-ray diffractors. For this, a preferably distinguished peak at low scattering angles is commonly utilized. When determining the particle size either from the specific surface area or with the help of the Scherrer equation, only a mean value for the particle size is obtained instead of the particle size distribution. On the other hand, the mean values found do not depend upon the dispersion of the pigments in a test medium since the measurements are performed on the pigment powders themselves.

Similar pigments or different batches of one and the same pigment can differ in mean particle size as well as in their particle size distributions. When pigments are "anisotropic"[1], that is not spherical, but, for example rod-shaped or platy instead, it would be necessary to characterize them by measuring the characteristic distributions of their lengths, their widths and their heights, etc. This, however, is only possible from tedious evaluations of electron microscopic or light microscopic photographs.

In the case of "isotropic" pigments or fillers, particle sizes are best described as logarithmic normal distributions. Logarithmic normal distributions are typical for properties that are limited by a value of zero to the lower end and in which the random variables that lead to the selected property concur in a multiplicate manner. That is to say that the change of a parameter is proportional to its value at any time. Apparently, this is true in the case of many [2] if not most pigments.

[1] Anisotropy relates to properties that are direction dependent.

The cumulative sum curve of the logarithmic normal distribution has the general form:

Equation 2.3

$$H_{(d)} = \frac{1}{\sqrt{2\pi} \cdot \ln\sigma} \cdot \int_0^d \frac{1}{d} \cdot e^{-\frac{1}{2}\left(\frac{\ln d - \ln\mu}{\ln\sigma}\right)} \cdot dd$$

$H_{(d)}$ is the cumulative occurrence of the feature d, μ the median value and σ the standard deviation of the distribution. The median is based on weight and, in this case, denotes the particle size that separates the weight distribution curve in half. The sum of all particles of a distribution with a particle size smaller than the median and the sum of all particles that are larger than the median have the same mass. Since larger particles are heavier than smaller particles, the median is larger than the arithmetic mean which is gained by dividing the total mass of all particles by their number. Since the signals of common particle size measurements are related to particle mass, the median is normally generated automatically. (If the measurement were to depend upon the number of particles, the arithmetic mean would be gained instead).

The distribution Equation 2.3 is called the "cumulative sum curve" since the sum of the incidences $H_{(d)}$ is plotted for increasing values of d. It has the general form

Equation 2.4

$$H_{(d)} = \sum_{k=1}^{j} d_k$$

and yields the mass fraction of all the particles that are smaller than d.

A different presentation of the log-normal distribution comes into being by differentiation of Equation 2.3 , whereby the so called "probability density function" is formed.

Equation 2.5

$$h_{(d)} = \frac{1}{\sqrt{2\pi}\ln\sigma} \cdot e^{-\frac{1}{2}\left(\frac{\ln d - \ln\mu}{\ln\sigma}\right)^2}$$

When plotting $h_{(d)}$, the frequency of a diameter d, against $\ln d$, a bell shaped curve is formed with a peak at $\ln\mu$ and inflection points at μ/σ $(= \ln\mu - \ln\sigma)$ and $\mu \cdot \sigma$ $(= \ln\mu + \ln\sigma)$, respectively.

It is easy to check if the particle size distribution follows a log-normal law (Equation 2.3) by simply fitting $H_{(d)}$-values against the respective d-values in a log-normal probability chart. In this chart, the x-axis is divided in a logarithmic scale whereas the y-axis represents the "Gauss integral". In case the particle size distribution does fit a log-normal law, the graphical representation yields a straight line. The intersection with the 50 % line represents the median (geometric mean), whereas the intercept with the 84.13 % and 15.87 % line, respectively, indicate the inflection points of the bell shaped probability function (Equation 2.5). Given that

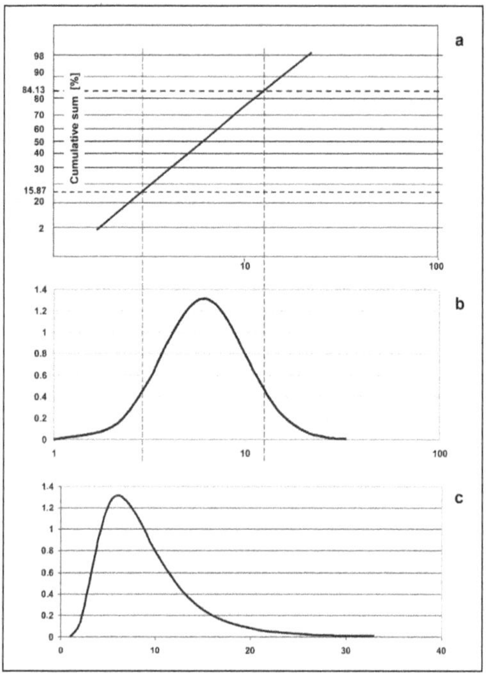

Figure 2.3: Log-normal distribution
a) linearization of the cumulative distribution function (Equation 2.3) in a log-normal probability chart, b) probability density function (Equation 2.5) of a log-normal distribution in a semi-logarithmic plot, c) probability density function (Equation 2.5) of a log-normal distribution in a linear plot

the particle size distribution follows a log-normal mode, 68 % of all diameters lie in between these two inflection points.

Figure 2.3 illustrates the connection between the cumulative distribution function in the log-normal probability chart and the probability density function itself. Furthermore, Figure 2.3 shows the shape of the probability density distribution function if the frequency of the diameters is plotted against the particle size instead of the logarithm of the particle size. Then the probability density function is no longer bell shaped but skewed to the left instead.

When a log-normal distribution exists and when the median and the standard deviation are known, then the mean particle size $\bar{\chi}$ and the most frequent particle size D in the linear plot can be calculated using Equations 2.6 and 2.7.

Equation 2.6 $$\bar{\chi} = anti\ lg\ (\mu + 1.1513 \cdot \sigma^2)$$

Equation 2.7 $$D = anti\ lg\ (\mu - 2.3026 \cdot \sigma^2)$$

For increasing standard deviations, yet at a given median, the difference between the most frequent particle size and the median increases twice as fast as the difference between the mean value and the median. The distribution therefore becomes increasingly broader and more skewed.

In theory, a preferably narrow particle size distribution of pigments is desirable. Furthermore, pigments with a narrow particle size distribution are usually easier to disperse since they have the tendency to form agglomerates that are more loosely packed. On the other hand, wide particle size distributions are often caused by coarse particle fractions that are often hardly dispersible at all. (These impurities are sometimes called “grit”.) For these reasons, it makes sense to describe particle size distribu-

Table 2.1: Properties of the pigments investigated (compare Table 2.2 and Figure 2.4)

Pigment	Production	Surface treatment	TiO_2-content in [%]	Relative scattering power in [%]	BET surface area in [m²/g]
1	Sulfate	Al_2O_3/SiO_2	92	105	9.4
2	Sulfate	Al_2O_3/SiO_2	95	112	12.9
3	Sulfate	Al_2O_3/org.	94	108	13.4
4	Sulfate	Al_2O_3/ZnO/org.	95	109	13.8
5	Chloride	Al_2O_3/SiO_2/org.	95	110	9.4

tions with the help of figures that reflect the medians of the particle sizes as well as the spread of particle diameters. The skewness [3] is such a measure.

Equation 2.8 $$S = \frac{3\,(\bar{\chi} - \mu)}{\sigma}$$

For a given standard deviation of a log-normal distribution, the skewness is larger, the farther the arithmetic mean and the median are separated from one another. In many cases the skewness of the pigment diameter distribution can be put in relation to paint properties [4]. Table 2.1 lists five titanium dioxide pigments in the rutile modification that are dispersed in four different binder systems. The particle diameter distributions in the cured paint films were measured from scanning electron microscopic (SEM) pictures with an automatic image analysis instrument. Agglomerates were detected as larger particles. The median values as well as the Standard

Table 2.2: Comparison of the skewnesses of titanium dioxide pigment particle size distributions in coatings and respective hiding power values (see Figure 2.4)

Resin	Pigment number	Skewness	Film thickness for ΔL = 2 in µm
A	1	0.1	50.6
A	2	0.0656	51.6
A	3	0.0653	45.6
A	4	0.0531	42
A	5	0.0698	44
B	1	0.077	55.3
B	2	0.0684	47.6
B	3	0.0811	47.6
B	4	0.0557	46
B	5	0.0609	46
C	1	0.112	65.6
C	2	0.1112	62
C	3	0.0812	51
C	4	n.b.	n.b
C	5	0.0807	55.3
D	1	0.1658	91
D	2	0.126	77.6
D	3	0.1357	71.3
D	4	n.b.	n.b
D	5	0.1776	83.3

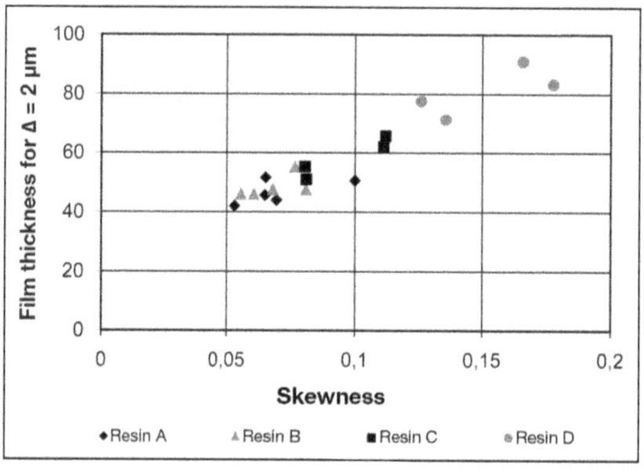

Figure 2.4: Hiding powers of coatings pigmented with five different titanium dioxide pigments in four different resins (A to D) as a function of the skewnesses of the pigment particle distributions in the coatings

deviations of the particle diameter distributions were determined with the help of probability charts (Figure 2.3a). From the data of these plots, the skewness values of the particle size distributions were calculated (Equation. 2.8). Parallel to that, the hiding powers of the paint films were determined following DIN 55 987 by applying the paints onto black and white contrast boards and measuring the dry film thickness at which a difference in lightness ΔL^* of 2 was reached. The hiding power of a coating is lower the higher the film thickness at this point is. Results are listed in Table 2.2. It can be seen that pigment #1 has comparatively low hiding power not only because of a low titanium dioxide content (i.e. much inorganic surface treatment), but also due to a comparatively poor distribution within the coating.

Figure 2.4 shows a plot of the film thicknesses for $\Delta L^* = 2$ versus the skewnesses of the respective pigment particle diameter distributions. It is clearly seen that the particle distributions have a large influence on the hiding powers achieved. Not only does the pigment itself have a grave influence, but also the resin system. This can be attributed to differences in the wetting behaviour of the resins as well as to their different abilities for stabilizing the pigments against flocculation. A further conclusion that can be made from these experiments is that, even when the particle size distributions of pigments become worse due to poor dispersion of by flocculation, their diameter distributions may still be described as log-normal functions.

2.3 Interactions between pigment particles

In principle, small particles of any nature undergo the same types of physical interactions that play a role between atoms and/or molecules. These interactions are

- dipole-dipole interaction
- induced dipole interaction
- London-van der Waals interaction (dispersive interaction)
- hydrogen bonds

These interactions were introduced in Chapter 1 of this book. Noteworthy was the large dependency of the interactions upon the separation distance. The interaction energies (energies of attraction) between atoms and/or molecules decline to the power of six regarding the separation distance, and interaction forces even to the power of seven. This might lead to the assumption that the interaction between particles depend solely upon the surface sited atoms of a particle. Although this supposition is correct in the case of dipole-dipole interactions, induced dipole interactions as well as hydrogen bonds, this is not true for the attraction resulting from London-van der Waals forces. When regarding dispersive interactions, it must be taken into consideration that every atom in one colloidal particle may undergo London-van der Waals interaction with every atom in the next particle. The sum of all of these very small interaction forces causes a considerable dispersive attraction between the pigment particles, which is by far greater than interactions from all of the other attractive mechanisms.

2.4 Van der Waals attraction between particles

Dispersive interactions between colloidal particles were investigated and treated mathematically by *Hugo Christiaan Hamaker* [5] (1905–1993). Apart from other cases, Hamaker developed the formulae that describe the interaction between two alike particles with diameters d_1 and d_2, respectively, in dependency of their surface separation distance a.

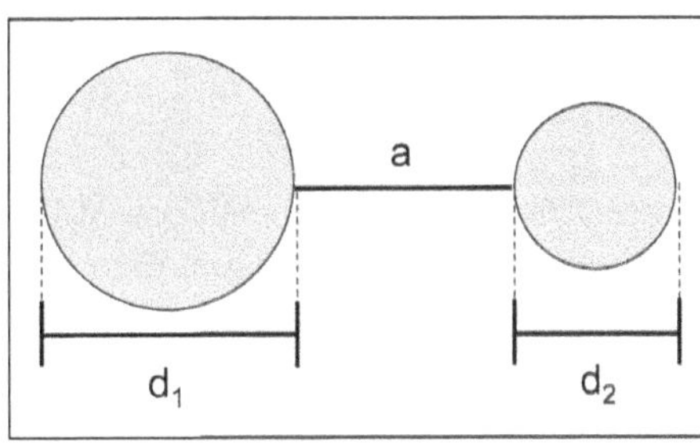

Figure 2.5: Schematic diagram to Equation 2.12

This situation is depicted in Figure 2.5. Hamaker concluded that the attractive energy V_{attr} can be expressed as the product between an energetic term A and a geometric term $H_{(d1,d2,a)}$ that depends upon the two diameters and the separation distance.

Equation 2.9 $$V_{attr} = -\frac{A}{12} \cdot H_{(d_1,d_2,a)}$$

The energetic term A in Equation 2.9 is known as the Hamaker constant. The Hamaker constant, and therewith the particle-particle interaction, becomes larger, the larger the number of atoms or molecules per volume within the particles are (molecule density q) and the larger their polarizabilities α and the characteristic frequencies v_0 of the electrons are.

Equation 2.10 $$A = \pi^2 \cdot q^2 \cdot \frac{3}{4} h v_0 \alpha^2$$

The molecule density may be determined from Avogadro's number N_A, the molecular weight M and the specific density ρ of the particles.

Equation 2.11 $$q = \frac{N_L \cdot \rho}{M}$$

The expression $$\frac{3}{4} h v_0 \alpha^2$$

in Equation 2.10 is the London-van der Waals constant that was already introduced in Chapter 1 to calculate the dispersive interactions between simple molecules. This shows that the formalism for describing dispersive interactions between colloidal particles and between simple atoms and/or molecules is very similar.

The geometric term $H_{(d1,d2,a)}$ in Equation 2.9 looks somewhat more complicated. According to Hamaker, it is:

Equation 2.12

$$H_{(d_1,d_2,a)} = \frac{y}{x^2 + xy + x} + \frac{y}{x^2 + xy + x + y} + 2ln\,\frac{x^2 + xy + x}{x^2 + xy + x + y}$$

whereby $x = \frac{a}{d_1}$ and $y = \frac{d_2}{d_1}$

Figure 2.6 shows the dependency of the geometric term $H_{(d_1,d_2,a)}$ for two particles with equal sizes of 1 µm diameter from the surface separation distance a. Beginning with very large values, the separation dependent expression loses in magnitude quickly with increasing separation distances. At a distance of 0.2 µm, equivalent to merely 20 % of the particle diameter, the geometric term has dropped practically to zero.

Inserting Equation 2.12, the expression for $H_{(d_1,d_2,a)}$, into Equation 2.9 yields the attractive energy between two spherical particles of the same chemical nature and therefore identical Hamaker constants in vacuum or, sufficiently accurate, in air, respectively.

Equation 2.13

$$V_{attr} = -\frac{A}{12} \cdot \left[\frac{y}{x^2 + xy + x} + \frac{y}{x^2 + xy + x + y} + 2ln\,\frac{x^2 + xy + x}{x^2 + xy + x + y}\right]$$

In other words, the attractive energy in dependence of the separation distance of colloidal particles is simply the surface separation function expanded by the multiple of *A/12*. Just as in the case of atoms or molecules, the negative sign in Equation 2.13 depicts the energy as an energy of attraction. Since the attraction energy arises from the dispersive interactions of every atom in the first particle with every atom in the second particle and vice versa, it is obvious that the attractive energy becomes larger with increasing particle size. Larger colloidal particles

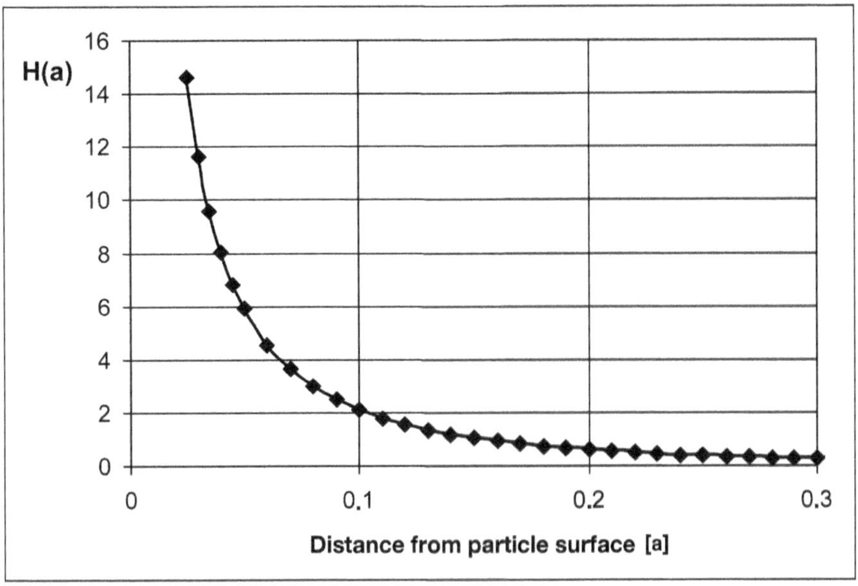

Figure 2.6: Graphical representation of Equation 2.12 (= geometric term in Equation 2.13) for $d_1 = d_2 = 1\ \mu m$

attract each other significantly stronger than smaller particles. This is easily verified by inserting different values for the diameters d_1 and d_2 into Equation 2.13 and calculating V_{attr}[2].

Within agglomerates, the individual particles are separated by only very small distances. Taking this circumstance into account, the contribution of the last two expressions in Equation 2.12 become negligible in comparison to the first term. This is verified either by a limit value evaluation or by inserting discrete values, e.g. $x = 0.01$ and $y = 2$, into Equation 2.12 and calculating the parts individually. Likewise, x^2 in the first term can be neglected in comparison to x, xy and y. Thus, Equation 2.13 simplifies to:

Equation 2.14
$$V_{attr} = -\frac{A}{12} \cdot \frac{y}{x(y+1)}$$

For two particles of uniform size, $y = 1$, and, since $x = a/d$, Equation 2.14 turns into:

Equation 2.15
$$V_{attr} = -\frac{A}{24} \cdot \frac{d}{a}$$

[2] Therefore, the proposition that nano-scaled particles have higher attraction forces than larger particles, as stated in many papers, is clearly false!

By differentiation of V_{attr} with respect to the separation distance, an expression for the attractive force F_{attr} between the particles is found.

Equation 2.16 $$F_{attr} = \frac{A}{24} \cdot \frac{d}{a^2}$$

Due to the mathematical formalism, the attractive force has a positive sign. Equation 2.16 indicates that the attractive force between the particles at a given separation distance increases with growing particle sizes or decreases, respectively, when the particles become smaller. Contrarily, the gravitational force F_{grav} acting on the particles is proportional to the third power of the particle diameter. With the aid of the gravitational constant g ($g = 9.80665\ m/s^2$) and the density ρ of the particles, the gravitational force can also be calculated.

Equation 2.17 $$F_{grav} = \frac{1}{6}\, \pi \cdot \rho \cdot g \cdot d^3$$

The different dependencies of the van der Waals attractive force and the gravitational force on the particle size have the consequence that, in the case of very small particles, the attractive forces amongst the particles become larger than the gravity acting on the single particles. This is the reason why they agglomerate. Were that not the case, life on earth would not be possible since dust would cloud the sun and creatures could not breathe.

In Figure 2.7, the van der Waals attractive forces on titanium dioxide particles [6] ($A = 2.68 \cdot 10^{-19}$ Joule, $\rho = 4.0\ g/cm^3$) at a distance of 10 nm are compared to the gravitational force acting on them. In the double logarithmic presentation, both force curves linearize so that linear fits may be drawn. Figure 2.7 shows that, under these conditions, the attraction forces between spherical titanium dioxide particles prevail over the gravitational forces as long as they are smaller than 100 µm. At a particle size of 0.3 µm, a characteristic value for commercial titanium dioxide pigments [7], the forces of attraction between the particles is approximately five powers of ten higher than the gravitational force acting on them. This is commonly taken as the reason not only that pigments agglomerate,

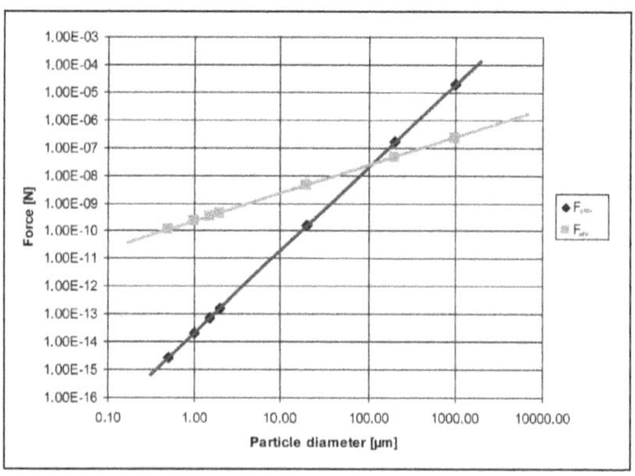

Figure 2.7: Comparison of gravity and the van der Waals attraction force of titanium dioxide particles in dependence of their size

but also why pigment powders flow in an agglomerated state, that is to say as agglomerates. By experience, bulk powders are only free flowing when they move as individual, single entities. In some cases this can be achieved by adding much smaller particles to the bulk material, especially when the bulk material particles have a particle size low enough to just favour particle-particle interaction over gravity [8]. The mechanism is explained by small particles coming in between the bulk powder particles, thereby enlarging their separation distances. According to Equation 2.16, enhancement of particle separation leads to a downward parallel translation of the F_{attr}-curve in Figure 2.17 to lower attractive forces so that the intersection with F_{grav}-curve is shifted to the left, that is to smaller particle sizes.

A further possibility to lower attractive forces between particles by physical means is by making their surfaces coarse. In that case, the embossed parts on the particle surfaces act as spacers. Thus, the surface treatment of titanium dioxide pigments (see Chapter 2.5) with high amounts of silica renders powders with improved flow properties, yet also more dusting. Methods for measuring the flow properties of cohesive powders are found in literature [9].

2.5 Surface treatment of pigments

Dispersing pigments and fillers sometimes proves to be an energy consuming process. Therefore, pigment producers take actions to offer easily dispersible products to the markets. In the case of inorganic pigments, inorganic surface treatments are state of the art. For that, the pigments are first subjected to intensive milling to ensure that possibly all agglomerates are turned into primary particles. In a second step, dissolved oxides of other components, such as aluminium, zirconium or silicium are added to aqueous suspensions of the milled pigments and, by adjusting the pH value towards neutral (pH 7), the deposition of the surface treating chemicals as hydroxides is initiated. Depending upon the execution of the precipitation reactions, mixed oxides such as aluminium silicate may cover the pigment surface. Other surface treatments ("post treatments") may consist of components of low solubility such as the phosphates of aluminium or titanium. Once again, it is London-van der Waals forces that attract the precipitated surface treatment chemicals and the pigment particles so that the pigments are completely ensheathed [10].

Subsequently, a further micronizing step follows. In the case of titanium dioxide pigments, a jet mill is used in which the grinding stock is fed tangentially along with steam into an annulus in which agglomerates and aggregates are micronized due to collisions with each other. Inorganic surface treatments may enhance the dispersibility, e.g. by coarsening the pigment surface so that the primary particles are kept from adjusting plane to plane (as already said above) or by reducing the effective Hamaker constant (see below). Pigments like titanium dioxide with a few per cent of silica (SiO_2) as a post treatment are very free flowing. Apart from improving dispersibility, inorganic surface treatments also determine the colloid-chemical properties of pigments. In special cases like titanium dioxide pigments, the surface treatments also suppress photoactivity [11].

Yet another possibility for improving the dispersibility of pigments is to apply organic surface treatments. These are employed by either simply spraying them onto the pigment powders, in some cases followed by mixing and milling, for example in a jet mill. The amount of organic substance applied depends upon both the specific surface area of the pigments as well as the intended pigment use. Today, glycolic substances such as trimethylolpropane or neopentyl glycole are commonly used. Silicone oils are often applied to pigments destined for usage in plastics. Surface treating with silicone oils leads to easily dispersible agglomerates, which is helpful especially in view of the lack of pigment stabilization mechanisms against flocculation in thermoplastic polymer melts (see Chapter 6.5.2).

Attractive forces between pigment particles come about due to London-van der Waals interactions. It is therefore self-evident that organic posttreatments improve their dispersibility by reducing the van der Waals interaction between them. This was investigated independent from one another by *Vervoorn* [12] and *Winkler* [13]. Both authors used an expanded Hamaker equation of *Vold* [14] to describe the influence of adsorbed molecules on the attraction energy of particles. Figure 2.8 shows the underlying model. Two spherical particles of equal radii R and a Hamaker constant A_p are covered with an adsorbed layer of thickness δ of a substance with a Hamaker constant A_s. The covered particles have a surface to surface separation distance of Δ in a medium with a Hamaker constant A_m. According to Vold, the attractive energy in this case is put together by the sum of separate products of Hamaker constants and geometrical terms H that depend upon the separation distance.

Equation 2.18

$$-12V = (A_m^{1/2} - A_s^{1/2}) \cdot H_s + (A_s^{1/2} - A_p^{1/2}) \cdot H_p + 2 \cdot (A_m^{1/2} - A_s^{1/2})(A_s^{1/2} - A_p^{1/2}) \cdot H_{ps}$$

The separation terms are given by:

Equation 2.19

$$H_{(x,y)} = \frac{y}{x^2 + xy + x} + \frac{y}{x^2 + xy + x + y} + 2 ln \frac{x^2 + xy + x}{x^2 + xy + x + y}$$

Whereas for H_s $\quad x = \frac{\Delta}{2(R + \delta)}$ and $y = 1$

for H_p $\quad x = \frac{\Delta + 2\delta}{2R}$ and $y = 1$

for H_{ps} $\quad x = \frac{\Delta + \delta}{2R}$ and $y = \frac{R + \delta}{R}$

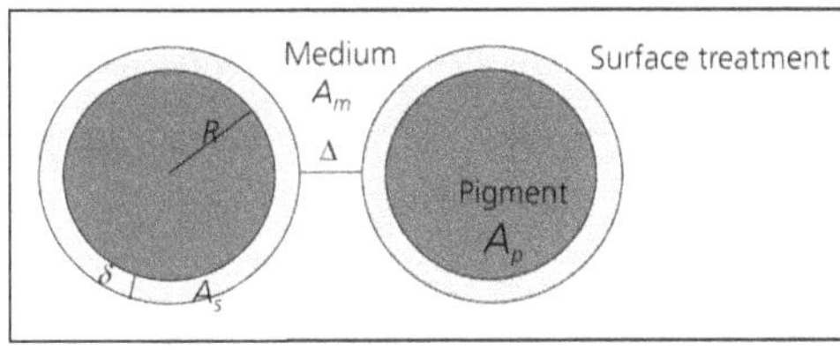

Figure 2.8: Principle model behind Equation 2.18

Vervoorn [12] interpreted the decline on the compressive strengths of titanium dioxide pigment agglomerates with increasing moisture contents by the influence of the water molecules on the van der Waals interaction. *Winkler*[13] coated a titanium dioxide pigment lacking an inorganic surface treatment with 27 different organic surface treatments. Both low molecular weight- as well as polymeric substances were employed in the study. The Hamaker constants of these reagents were determined (Equation 2.10) wherefore their polarizabilities were ascertained via the Clausius-Mosotti equation (Equation 1.16) and the characteristic frequencies were gained using the Cauchy dispersion formula (Equation 1.23). Surface coverages of the pigment samples with the organic substances were determined by carbon analysis. They laid between 0.1 and 5.3 Ångstroms (1 Ångstrom = 10^{-10} m. A carbon-carbon bond in a hydrocarbon is approximately 1 Ångstrom), whereby the volatility of the organics in the jet mill had an influence on their retention. Polymeric surface coating materials therefore led to higher loads. The mean particle to particle distances were estimated from apparent powder density measurements and were normalized to the particle separation distance in the untreated reference. In this way, energies of attraction between the particles could be calculated using Equation 2.18, whereby the mean separation distance in the reference sample was set to one nanometre.

An apparatus was developed with which agglomerate diameters could be determined to an accuracy of ± 2 µm and with which the work that was necessary to crush them could be measured simultaneously. The quotient between this work and the corresponding agglomerate volume was taken as a measure for their compression strength. Sieve fractions of agglomerates of the treated pigments were generated and measurements were performed on 60 individual agglomerates, each of about 100 nm (0.1 mm) diameter. Compression strength distributions were gained from the data that could be expressed as log-normal distributions (Equation 2.3 and 2.5). Table 2.3 lists the organic substances used, their retained amount on the pigments, the medians of the compression strengths as well as the compressibilities of the agglomerates. Compressibilities reflect the extent of agglomerate distortion upon breakage[3]. Substances 20 to 22 are special in the sense that they are polyester-modified polysiloxanes with free terminal Si-H groups.

[3] The exact definition is presented in the original literature (J. Winkler, farbe+lack 94 (1988) 108).

Table 2.3: The influence of organic surface treatments on the compressive strengths $\tilde{D}_g$ and compressibilities χ of 100 nm agglomerates of titanium dioxide pigments

Substance No.	Weight-% retained	Surface treatment	$\tilde{D}_g$ [N · cm^{-2}]	χ
Blank	0		0.97	055
1	0.47	Tinuvin 1130*	0.95	0.53
2	0.41	Tinuvin 292*	0.75	0.56
3	0.18	Dioctylphtalate	1.5	0.48
4	0.06	N-(3-Disobutylamine-propylmorpholine)	0.71	0.54
5	0.17	3-(2-ethyl-hexoxi)-propylamine-(1)	0.52	0.5
6	0.02	1.3-Dimethyl-5-tert.butyl benzene (5-tert.butyl-metaxylole)	0.77	0.66
7	0.01	2,4-Dimethyl-3-dimethylamino pentane	0.77	0.63
8	0.09	3-Isobutoxi-propylamine-(1)	0.77	0.57
9	0.09	3-n-Butoxi-propylamine-(1)	0.85	0.61
10	0.31	Triethanolamine	0.31	0.73
11	0.01	2,4-Dimethyl-pentanol-(3)	1.08	0.62
12	0.05	3-Ethoxi-propylamine-(1)	0.88	0.56
13	0.02	5-Methyl-isoxazole	0.58	0.62
14	0.01	1,3-Dioxolane (Glykole methylene-ether)	0.8	0.61
15	0.33	Polyethylenoxide, Mol.-weight appr. 200	0.45	0.65
16	0.41	Polydimethylsiloxane	0.02	0.91
17	0.33	lightly branched silicone oil with alkylhydroxy group endings	0.19	0.65
18	0.33	Polyestermodifiied polysiloxane	0.65	0.71
19	0.39	Polyestermodifiied polysiloxane	0.09	0.79
20	0.54	Polyestermodified polysiloxane with 20 mol % rest reactivity; 40 Si-O units	1.53	0.62
21	0.41	Polyestermodified polysiloxane with 20 mol % rest reactivity; 20 Si-O units	1.35	0.62
22	0.52	Polyestermodified polysiloxane with 20 mol % rest reactivity; 15 Si-O units	1.47	0.57
23	0.44	Polyestermodifiied polysiloxane	0.44	0.65

* For chemical structure see www.chemblink.com

Continue Table 2.3:

Substance No.	Weight-% retained	Surface treatment	$\tilde{D}_g$ [N · cm^{-2}]	χ
24	0.16	Polyestermodifiied polysiloxane	0.45	0.69
25	0.32	Polyestermodifiied polysiloxane	0.46	0.7
26	0.4	Polyestermodifiied polysiloxane	0.65	0.75
27	0.41	Polyestermodifiied polysiloxane	0.38	0.78

To start with, the data show that organic surface treatments have a large influence on agglomerate strength. Compressive strengths ranged between 1.35 N/cm² (uncoated reference sample) to 0.02 N/cm² (in the case of a treatment with pure polydimethylsiloxane, sample #16). The samples 20 to 22 have even higher compression strengths than the reference sample which is probably the result of a hydrolysis of Si-H groups to silanol groups under the conditions of steam milling and a subsequent reaction of the generated Si-OH groups with the Ti-OH groups on the surface of the particles to form Ti-O-Si bonds. In case the polysiloxane molecules react this way with more than one TiO_2 particle, the organic molecules would form bridges that might have a strengthening influence in agglomerates. In any case, these samples do not follow the trend of all the other samples with respect to many properties.

Figure 2.9 is a plot of agglomerate compressibilities vs. agglomerate strengths. Compressibilities of harder agglomerates tend to be lower than those of softer ones. The higher the compressive strengths are, the more densely they are packed.

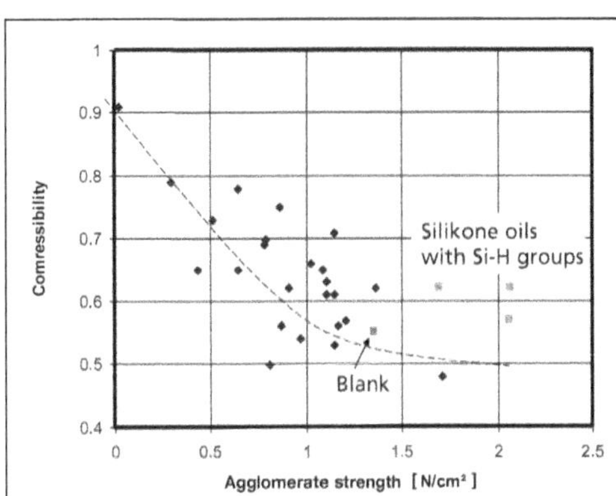

Figure 2.9: Compressibility of pigment agglomerates in dependency of their compressive strengths

In Figure 2.10 the fineness of grind readings (DIN EN 21 524 or ISO 1524) of the pigments after dispersion in an alkyd resin formulation are plotted against their respective agglomerate strengths. A higher fineness of grind is an indication of poor dispersion. As expected, harder agglomerates are more difficult to disperse and the comparison with Figure 2.8 shows that agglomerates which are packed more densely are more difficult to disperse. This

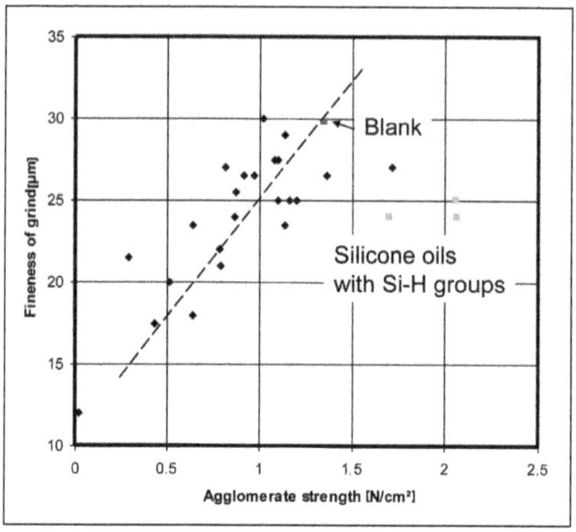

Figure 2.10: Fineness of grind (Hegman gauge) of titanium dioxide pigments tested in a standard dispersibiltiy test in dependency of agglomerate strength

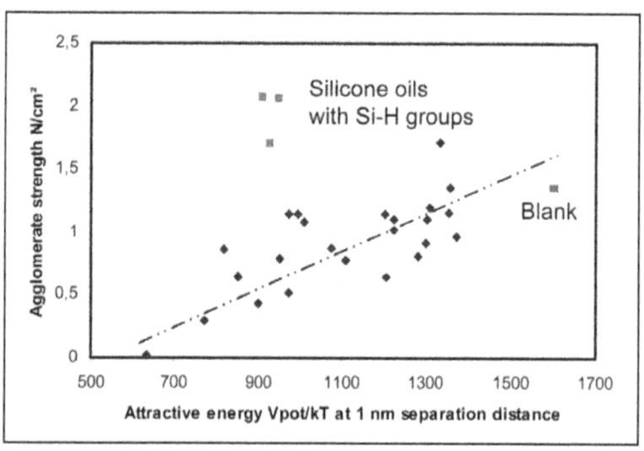

Figure 2.11: Comparison of measured agglomerate strengths and calculated energies of attraction for titanium dioxide pigments of 0.3 μm size

makes sense, since the primary particles in the more densely packed agglomerates have more contact points amongst each other which lead to stronger adhesive forces. The reactive silane modified samples 20 to 22 don't behave in terms of dispersibility in the manner one might first expect from their compressive strengths. This could be because the bridging silane molecules attach to only two or more primary particles which, in the case of 0.3 μm particles, do not necessarily effect the grindometer fineness. On the other hand, bridging chemical bonds might provoke an additional hindrance for the pigments particles to rearrange during compression so the compressive strengths could increase anyway.

Figure 2.11 is a plot of the logarithmic mean compressive strengths against attractive energies calculated for the pigment samples using Equation 2.18. In this case, the attractive energies are reported as multiples of $k \cdot T$ (k = Boltzmann constant, T = 293 K). With the exception of the samples 20 to 22 treated with reactive silanes, there is a correlation between the calculated attractive energies and the measured mean compressive strengths. This indicates that the improvement of dispersibilities of pigments by organic surface treatments is primarily caused by their influence on the dispersive Londen-van der Waals interactions between the individual particles. The lower the Hamaker constant of the organic treatment is, the lower are also the attractive forces between the particles at the time the agglomerates are formed. This leads to less densely packed agglomerates with a smaller number of contact sites of the particles and lower attractive force per contact site.

Finally, Figure 2.12 shows the relationship between the dispersibility of the pigment samples and the calculated attractive energies. The dependency shown is also accessible from Figures 2.10 and 2.11.

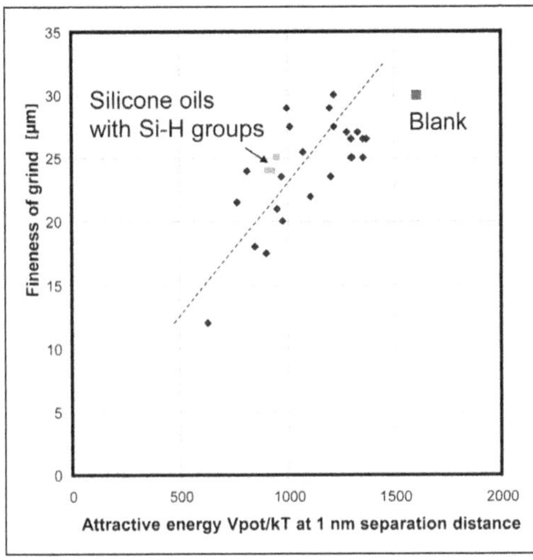

Figure 2.12: Fineness of grind (Hegman gauge) of titanium dioxide pigments in a standard dispersibility test in dependency of the calculated attractive energies between the particles of size 0.3

Whereas easily dispersible pigments are normally favoured, smaller dispersive interactions between pigment particles unfortunately also lead to pigments that show more dusting while handling them. After all, the precondition for the generation of dust is that individual pigment particles readily disconnect from the rest of the agglomerates and that they remain suspended in air for a longer period of time because of their low weight. The tendency to create dust is easily measured, for example, by letting a given amount of pigment powder fall into a glass cylinder, followed by the application of a defined stream of air crossing through the apparatus. The dust is then carried away and may be captured in a filter and weighed. Figure 2.13 shows such an apparatus. The percentage of material collected in the filter based on the pigment mass initially placed into the trap door at the top of the cylinder may be taken as measure for dusting. Figure 2.14 confirms that pigments with less dispersive attraction generate more dust.

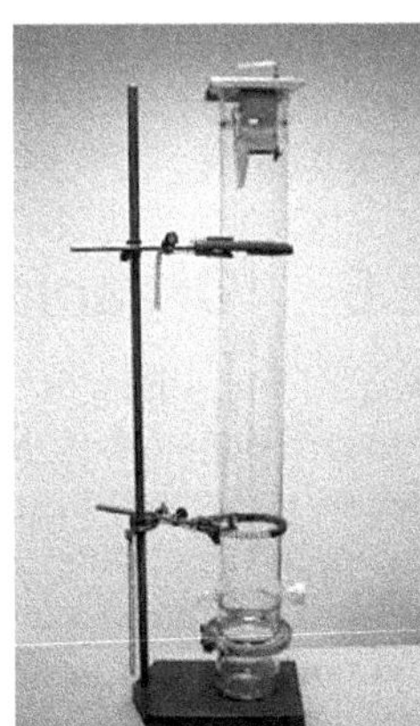

Figure 2.13: Apparatus for determining dusting behaviour of pigments
Source: Sachtleben Chemie GmbH

Organic surface treatments with low Hamaker constants lower the agglomerate strengths, thereby improving their dispersibility. A disadvantage is, however, that they undergo only little interaction with other molecules. The polydimethysiloxane on pigment sample #16 of Table 2.3 which led to the lowest compressive strength, the best dispersibility (and the highest generation of dust) is commonly used as a releasing agent in injection moulding of thermoplastics. There it has the function to make it easier to extract the plastic part from the injection dye. On the other hand, such a silicone

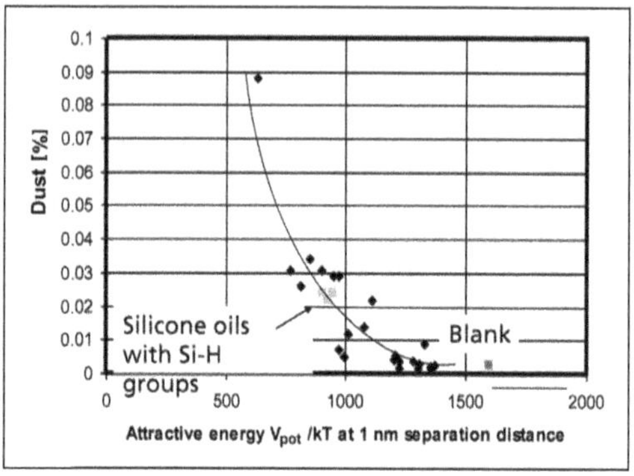

Figure 2.14: Relation between calculated energies of attraction between titanium dioxide pigment particles of 0.3 μm size and their dusting behaviour

oil, when on a pigment surface, prevents the adsorption of stabilizing resin and dispersing agent molecules on the pigment surface[4]. In this case, although the pigment is easily deagglomerated, it would flocculate dramatically in the medium. The titanium dioxide pigment #16 therefore had by far the poorest scattering power in an alkyd based formulation in which the fineness of grind was determined! If good dispersibility is to be combined with favourable colloid-chemical properties also, then the approach of using organic surface treatment chemicals with a higher Hamaker constant, yet in larger quantities, renders the best results. When the referred glycolic substances are used, then it is provided for that they desorb both in aqueous, as well as in solvent based formulations, thereby clearing the way for large, stabilizing molecules to get to the pigment surface. This is easily demonstrated by manufacturing thin-layer chromatography plates from the pigments on glass panels, if necessary by using a bit of gypsum as a binder. The organic treatments may be dappled onto the plates and can be displaced by using the solvents from the formulations as the mobile phase in a chromatographic experiment. If the surface treatments migrate with the solvents, then (even quantifiable) proof is gained that the surface treatments undergo adsorption/desorption equilibria in the solvent or solvent mixtures of the formulations.

2.6 Organic and inorganic pigments

Apart from fillers, organic as well as inorganic pigments are used in polymeric composite materials. Inorganic pigments are utilized not only to produce colour, but also by virtue of other properties such as antistatic and anticorrosive behaviour or to affect stone chipping performance. On the other hand, organic pigments are incorporated solely for adding colour to the composites. Pigments act as colorants by the interplay between the absorption and the scattering of light; two properties which are designated by the absorption coefficient and the scattering coefficient, respectively. The absorption coefficient rises with decreasing pigment

[4] In aqueous media, the presence of a polydimethylsiloxane could even prevent pigment wetting completely!

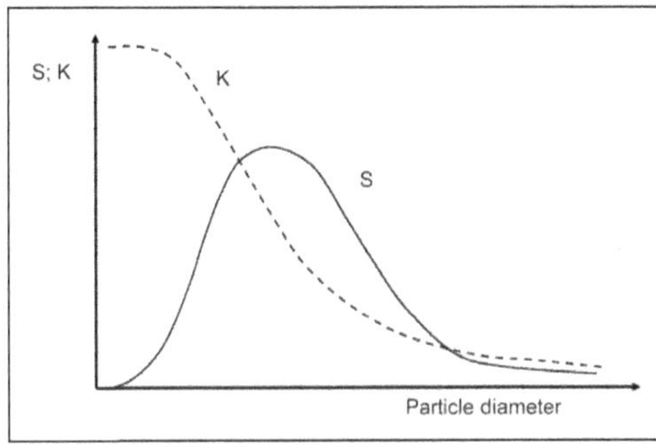

Figure 2.15: Dependencies of the light absorption coefficient K and the light scattering coefficient S on pigment particle size (principle scheme)

particle size and is directly proportional to the pigment loading in a formulation. The scattering coefficient is largest for particles whose size (diameter in the case of spherical particles) corresponds to about half the wavelength of visible light. Both smaller as well as larger particles are less efficient in terms of light scattering. Next, light scattering ability is best, the larger the differences in the refractive indices of the pigment and the surrounding media are. With increasing pigment load, the ability to scatter light first becomes larger, but then goes through a maximum value before declining again (viz. Figure 2.15).

Normally, inorganic pigments have a higher refractive index, so that they are superior in their ability to scatter light. Pigments that scatter light well are called "hiding pigments". These pigments do not need the presence of additional white pigments to impart colour to a composite material. Organic pigments tend to be of finer particle size, making them more "transparent". In comparison to inorganic pigments, they are quite intensely coloured. Detrimental aspects may be their lower heat stability, their (in most cases) poorer light fastness and the poorer weatherability. Organic pigments have the tendency to dissolve when moisture diffuses through composites due to condensing and evaporating moisture. They "migrate" and cause "chalking".

Typical representatives of inorganic pigments are oxidic compounds such as titanium dioxide (TiO_2) in the anatase or rutile crystal structure, the divers iron oxides with iron in different oxidation states, for example α-FeOOH (goethite), γ-FeOOH (lepidocrite), α-Fe_2O_3 (hematite), γ-Fe_2O_3 (maghemite) and Fe_3O_4 (magnetite), chromium oxide (Cr_2O_3), but also mixed metal oxide (MMO) pigments such as titanates etc. Carbon blacks are classified as inorganic pigments just like the sulfides and selenides of some metals like zinc (ZnS) and cadmium (CdS, CdSe).

Organic pigments are classified as azopigments, polycyclic pigments and anthraquinone pigments. Table 2.4 lists different types of products along with some of their most important properties.

Table 2.4: Organic pigments

	Fastness to weathering	Light fastness	Resistance to solvents	Migration fastness	Use in
Azopigments					
Monoazo yellow and orange pigments	reasonable	good	poor	poor	Paints, printing inks
Disazo pigments	reasonable	reasonable	reasonable	reasonable	Printing inks, plastics
Beta-Naphthole pigments	good	good	poor	poor	Paints
Naphthole AS pigments	good	good	poor	poor	Printing inks, paints
Laked azo pigments	poor	reasonable	reasonable	reasonable	Printing inks
Benzimidazolone pigments	good	very good	very good	very good	Automotive paints, printing inks, plastics
Condensed disazo pigments	good	good	good	good	Plastics, synthetic textiles
Metal complex pigments	good	good	good	good	Industrial paints, partly automotive paints
Isoindoline- und isoindolinone pigments	good	good	good	good	Plastics, high end paints
Polycyclic pigments					
Phthalocyanine pigments	good	very good	very good	very good	Industrial paints, automotive paints, plastics
Quinacridone pigments	good	very good	good	very good	Industrial paints, automotive paints, plastics
Perylene- und perinone pigments	very good	good	good	good	Industrial paints, automotive paints, special printing inks
Thioindigo pigments	good	good	good	good	Industrial paints, automotive repair paints

Continue Table 2.4:

	„Fastness to weathering“	„Light fastness“	„Resistance to solvents“	„Migration fastness“	Use in
Polycyclic pigments					
Diketopyrrolopyrrole pigments	very good	very good	very good	very good	Automotive paints
Dioxazine pigments	good	good	very good	very good	Paints, printing inks, synthetic textiles, plastics
Triarylcarbonium pigments	reasonable	reasonable	reasonable	reasonable	Printing inks
Quinophthalone pigments	reasonable	reasonable	reasonable	reasonable	Plastics, paints
Anthrachinon-Pigmente					
Anthrapyrimidin-Pigmente	very good	very good	good	good	Industrial paints, automotive repair paints
Flavanthron-Pigmente	very good	very good	good	good	Automotive paints
Pyranthron-Pigmente	good	good	good	good	Industrial paints, automotive repair paints
Anthranthron-Pigmente	very good	very good	good	good	Industrial paints, metallics for automotive use

Because of their commonly lower pigment particle size, organic pigments are frequently more difficult to disperse compared to inorganic pigments, which tend to be coarser. This is predominantly attributed to particle size, since both transparent inorganic pigments as well as carbon blacks require comparable efforts for dispersion[5]. Since most inorganic pigments have polar surfaces, they are normally easier to wet. This is also due to mostly smaller void volumes within the agglomerates of inorganic pigments, which further enhance wetting (see Chapter 3.3). Most organic pigments are considerably more expensive than inorganic pigments and, the better their quality is, the more expensive they become. For economic reasons it is therefore mandatory that they are thoroughly dispersed. When deagglomeration is poor, or when flocculation takes place, 30 to 50 % of the tinting strength potential may easily be lost.

As to the properties and numerous uses of fillers in coatings and other composite materials, reference is made to the literature [17].

2.7 Literature

[1] DIN 66 132

[2] G. Kämpf, W. Liehr, H. G. Völz, farbe + lack 76 (1970) 1105; F. Kindervater, farbe + lack 71 (1965) 445; G. Kämpf, farbe + lack 71 (1965) 353

[3] L. Sachs, "Angewandte Statistik", 6th Edition, Springer Verlag, Berlin 1984, p. 81

[4] J. Winkler, farbe + lack, 89, (1983) 332

[5] H. C. Hamaker, Physica 4 (1937) 1058

[6] J. Winkler, farbe + lack, 94 (1988) 263–270

[7] J. Winkler "Titanium Dioxide", U. Zorll (Editor), Vincentz Verlag (2003), p. 55

[8] H. Rumpf, Chemie-Ing.-Techn., 46 (1974) 1

[9] D. Schulze, Chem. Ing. Techn. 67 (1995) 60–68

[10] J. Winkler, "Titanium Dioxide", U. Zorll (Editor) Vincentz Verlag, Hannover (2003), p. 38

[11] J. Winkler, "Titanium Dioxide", U. Zorll (Editor) Vincentz Verlag, Hannover 2003, p. 71–78]

[12] P.M.M. Vervoorn, Colloids and Surfaces, 25 (1987) 145–154

[13] J. Winkler, farbe + lack, 94 (1988) p. 108–114 und p. 263–269

[14] M. Vold, J. Colloid Sci., 16 (1961) 1–12

[15] Industrial Inorganic Pigments, G. Buxbaum Ed., VCH, Weinheim, 1993

[16] W. Herbst, K. Hunger, Industrielle Organische Pigmente, VCH, Weinheim, 1987

[17] D. Gysau, Fillers for Paints, 2nd Edition, Vincentz Verlag, Hanover 2011

[5] The reason for this is discussed in Chapter 5, namely that smaller agglomerates are more difficult to stress.

3 Wetting of pigment surfaces

Typical values for the specific surface areas of pigments and fillers lay between <1 m^2/g up to about 300 m^2/g. When pigments are introduced into liquid media, all of this surface must be wetted completely.

In this chapter, principles of the physical chemistry of wetting processes are first treated for the case of large, flat surfaces. After this, the specific aspects of the wetting of pigments and fillers are discussed.

3.1 Relevancy of wetting for the dispersion process

It was shown in Chapter 2 that, due to the van der Waals attraction, dry pigment powders are always in an agglomerated state. While being dispersed, the agglomerates are disintegrated into the primary particles because of mechanical stresses that are inflicted onto them. The forces that are necessary for that to happen are lower if the agglomerates are wetted. The reason for this is that the effective Hamaker constant A_{eff} which is responsible for the attraction of the particles (see Equation 2.13), is lower in a liquid medium than in air. For a particle with a Hamaker constant of A_P in a medium having a Hamaker constant of A_M, the effective Hamaker constant is

Equation 3.1 $$A_{eff} = (A_P^{1/2} - A_M^{1/2})^2$$

Equation 3.1 demonstrates that the effective Hamaker constant, and thereby the dispersive interaction between the pigment particles, becomes smaller, the more similar A_P and A_M become. This means that the interaction between the pigment and medium become more and more predominant compared to the interaction the particles undergo amongst each other.

A titanium dioxide pigment in the rutile modification has a Hamaker constant of approximately $2.7 \cdot 10^{-19}$ joule [1]. Air has a Hamaker constant of merely $3.8 \cdot 10^{-25}$ joules, whereas water lies by $8 \cdot 10^{-20}$ joules and organic liquids between $5 \cdot 10^{-20}$ and $8 \cdot 10^{-20}$ joules. In an agglomerate completely wetted with water, the van der Waals attraction is, according to Equation 3.1, merely about 20 % of that of a dry

Jochen Winkler: Dispersing Pigments and Fillers

ISBN 978-3-86630-852-7

agglomerate. For this reason, pigment wetting facilitates the dispersion process considerably. In former times, before the development of closed, high speed bead mills allowed to disperse with high powers, it was customary to "pond" the pigments prior to dispersion. This meant that the pigments were stirred into a solvent or the "mill base"[1] at least a few hours before dispersion, if not even over night. The actual dispersion was carried out when all of the entrapped air had left the agglomerates, that is, when air bubbles no longer rose to the surface of the batch.

Today, one normally does not take the time so that it is possible that there are still unwetted pigment agglomerates present after the dispersion. When the final wetting process then takes place in the milled batch, its viscosity may rise to an extent that thickening happens.

Wetting, in a thermodynamic sense, is determined by the concurrence between the surface tensions of the liquid components, the free surface energies of the solids and the interfacial tensions between the liquid and the solid phases. The kinetics of wetting is further dependent upon the viscosity of the liquids involved. Wetting becomes more sluggish the higher the viscosity is.

3.2 Surface tension

Upon wetting, the interface solid/gas (or solid/vacuum) is replaced by the interface solid/liquid. When droplets of different liquids are brought to a flat solid surface, one can observe that they behave differently, depending upon the natures of the liquid and the solid surface. A drop of mercury on a glass plate looks like and rolls like a ball when the plate is tilted. On the other hand, a drop of water forms a calotte that looks like a split lentil. Finally, a drop of ethanol spreads spontaneously and completely with a "bow wave" on the outer fringe of the expanding droplet.

The reason for the different behaviours lies in the different interactions of the materials involved: substrate, liquid and gaseous phase.

An important physical quantity with which the differences in wetting by these three liquids in the experiments may be explained is the "free surface energy". The free surface energy is defined as the isothermal, reversible work that is necessary to create a surface or, in the case of the "interfacial tension", an interface. "Isothermal" means that the creation of the surface is carried out without any change in temperature. "Reversible" means that the change occurs while a thermodynamic equilibrium is maintained. This expresses that a liquid/gas or solid/liquid interface may either grow or shrink if the influencing variables change.

[1] Actually it should be named "dispersing base", since the pigments are not milled, but dispersed instead. Nevertheless, "milling" has become the established term in our trade.

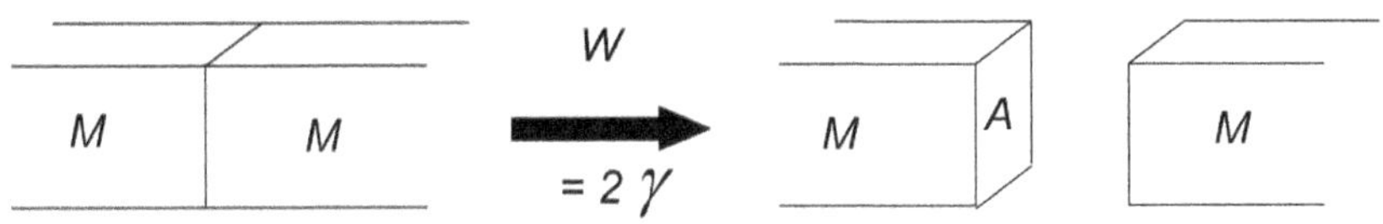

$$\frac{W}{2 \cdot A} = \frac{work}{area} = \frac{free\ energy}{area} \quad \frac{Nm}{m} \ or \ \frac{N}{m}$$

Figure 3.1: Free surface energy definition

Figure 3.1 depicts the meaning of the free surface energy by example of the generation of two surfaces of a substance. The substance is divided, whereby the work W needs to be delivered to the system in order for two surfaces of area A to be created. The free surface energy γ is the quotient between the work W and the total area $2 \cdot A$.

Equation 3.2 $$\gamma = \frac{W}{2 \cdot A}$$

After rearrangement, Equation 3.2 is useful for computing the work necessary to create any surface towards air, given that the free surface energy γ is known. The free surface energy has the physical dimension work/area, thus joules/square meter (J/m^2) or newton · meter/square meter (Nm/m^2). Cancelling the dimension "meter", leads to newton/meter (N/m). In the case of liquids, the free surface energy is normally called "surface tension". For practical reasons, surface tensions are mostly reported in the dimension milli-Newton per meter, thus thousandth of a newton per meter, which leads to the maybe deceptive expression of the physical dimension as "mN/m" [2]. The actual reason for the existence of free surface energies or surface tensions, respectively, lies in the fact that, whereas molecules within the bulk of a substance, for example a liquid, are surrounded completely by other molecules of the same kind, this is not true for those at the surface. All molecules undergo the interactions that were introduced in Chapter 1 of this book. Within a material, these physicals interactions are equally effective in all direction. Molecules on the surface, however, only have direct neighbours to interact with "alongside" and "underneath" them, whereas in the direction "above" them, to the gaseous phase for example, there are little or no interactions, alone because the concentration of molecules in the gaseous phase is low. Since an inward force pulls the molecules on the surface of liquids into the bulk, liquid surfaces act like strained, elastic membranes. Although incorrect[3] in a physical sense, yet ostensive, the surface

[2] In older literature, surface tensions are reported in the dimension dyne/cm (cgs system), which corresponds numerically with mN/m.

[3] A row of molecules is not one dimensional, but has a spatial extension.

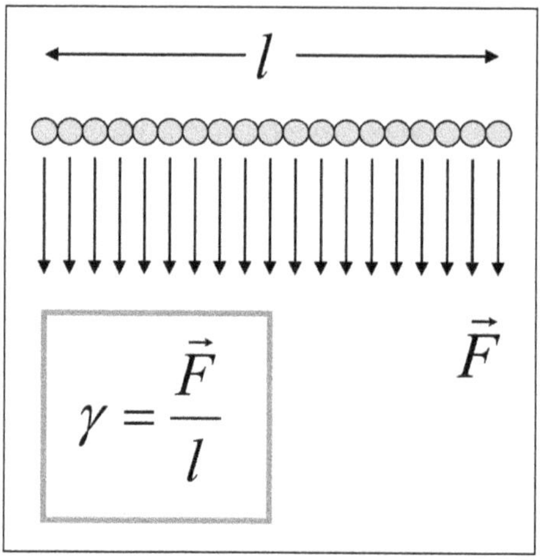

Figure 3.2: Alternative definition of free surface energy

tension is sometimes understood as the force acting on a row of adjacent molecules of the length *l*. Figure 3.2 visualizes this picture.

The surface tension of water is, for example, the reason why droplets of fog are spherical. The sphere is the geometric figure with the lowest ratio of surface to volume. The reason why small droplets have a higher vapour pressure than larger droplets may be explained by surface tension as well. The smaller droplets, i.e. the higher the curvature of their surfaces are, the fewer are the number of adjacent neighbouring molecules that pull the molecules on the surface into the droplets, so it is easier for them to pass over into the gaseous phase than from a larger droplet[4].

There are a number of ways to measure the surface tension of liquids. For determining so called static surface tensions, the Wilhelmy plate method, the Du-Nouy ring method and the capillary rise method are frequently employed. Next to these, bubble pressure tensiometers are used to measure dynamic surface tensions in aqueous solutions containing surfactants. This method is helpful for investigating the speed with which surface active molecules diffuse to the interface liquid/air. All of these methods are described in textbooks of physics, for example [2] or may easily be found on the internet.

Surface tensions of liquids decrease with increasing temperature due to the Brownian motion of the molecules that becomes more predominant, thus acting against the inward pull on the surface sited molecules. At the critical point, that is the temperature T_c, at which a gas can no longer be liquified by pressure increase, the surface tension becomes zero, as expressed by the Eötvös equation.

Equation 3.3 $$\gamma = \frac{k(T_c - T)}{\sqrt[3]{V_{mol}^2}}$$

In Equation 3.3, *k* represents a universal constant, independent of the type of liquid. It has a value of $2.1 \cdot 10^{-7}$ J/K $\cdot$ mol$^{-2/3}$. V_{mol} is the molar volume and *T* the temperature in Kelvin.

[4] Ostwald ripening, that is the growth of particle size in colloidal suspensions in the course of storage, may be explained in the same way.

When mixing two immiscible liquids (like oil and water) together, two separate "phases" are established so that an interface is created. The formation of two phases from mixtures of two liquids may be understood in that the separate types of molecules undergo substantially more physical interaction amongst each other than with the molecules of the other sort. At the interface that is created, there is no longer a surface tension but an "interfacial tension" at work. Analogously to the surface tension, the interfacial tension is the isothermal, reversible work that is required to form the interface. It is the interfacial tension, for example, that causes the oil droplets in a salad vinaigrette to become larger in the course of time and eventually separate from the vinegar. During the demixing, the interface between the oil phase and the aqueous phase becomes continuously smaller so that energy is set free.

Solids also have surface tensions. However, these are commonly (and correctly) termed "free surface energies".

When small amounts of immiscible molecules are added to liquids, these tend to accumulate on their surface. When this happens, the surface is made up of a layer of molecules that have less interaction with the bulk liquid. Therefore, the surface tension is lowered in comparison to that of the pure liquid. The reduction of the surface tension of water when soap or some organic molecules are added can be explained this way [3].

All in all, the factors surface tension, interfacial tension and free surface energy that are so important for wetting behaviour are all based upon the physical interactions between molecules or atoms, respectively.

3.3 Young equation

Thomas Young [4] described the wetting behaviour of a flat glass substrate by mercury, water and ethanol in the year 1805. He reported that a drop of "moderately clean" mercury has a contact angle of approximately 140°, whereas the contact angle that a drop of water forms on the glass plate is very small. The meaning of the expression "contact angle" of a drop of liquid is depicted in Figure 3.3. When viewing a droplet sideways, the contact angle can be measured by applying a tangent to the curvature of the drop at its base point.

Figure 3.3: Droplet of liquid with contact angle

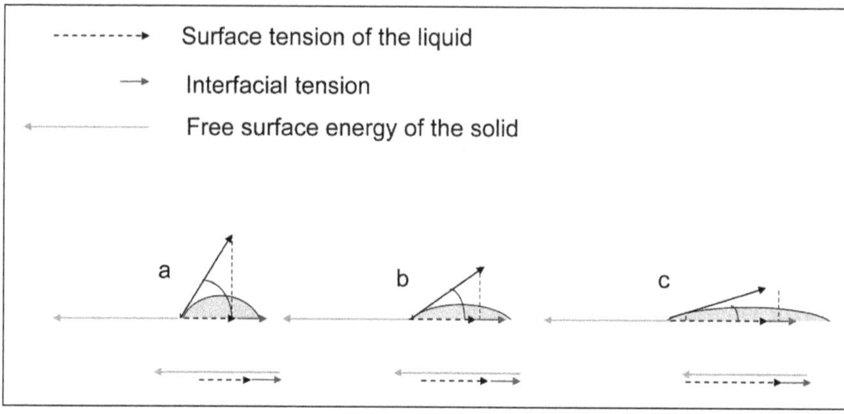

Figure 3.4: Flowing of a drop of liquid freshly applied to a solid surface; a: at the beginning; b: during deliquesce; c: at equilibrium

The contact angle comes forth as a result of the interplay of energies[5] (work) that must be put about or that are released when liquids come into contact with solid surfaces. It is the free surface energy of the solid γ_S that has a spreading influence on the droplet, whereas the projection of the vector of the surface tension of the liquid onto the base of the droplet and the interfacial tension both pull the droplet together. In thermodynamic equilibrium, Equation 3.4 holds:

Equation 3.4 $$\gamma_S = \gamma_{SL} + \gamma_L \cos\theta$$

Equation 3.4 is called the Young equation or, sometimes, the Young-Dupré equation. Figure 3.4 shows the interdependencies schematically for the example of a drop of liquid freshly applied to a solid surface. At the beginning (Figure 3.4a) the free surface energy of the solid is much larger than the sum of the interfacial tension and the projection of the vector of the surface tension of the liquid onto the base line of the droplet. Therefore, the droplet starts to spread so that the contact angle becomes smaller, thus increasing the influence of the surface tension of the liquid (Figure 3.4b). At equilibrium (Figure 3.4c) Equation 3.4 is satisfied.

Surface energies of solids in vacuum are much higher than when the solids are in contact to air which contains, for example, vaporized liquid molecules. The adsorption of such molecules not only lowers the free surface energy of the solid but the interfacial tension as well since molecules of liquid are already adsorbed to the surface. Frequently, adsorbed water molecules from moisture cause the surface energy to be lower. For this reason, Equation 3.4 should actually be corrected by insertion of the spreading pressure π. The spreading pressure is the difference

[5] Energy and work are equivalent quantities. Energy is defined as the ability to perform work.

between the free energy of the surface in vacuum γ_{SV} and its free energy in the presence of adsorbed molecules γ_S.

Equation 3.5 $$\gamma_{SV} - \gamma_S = \pi$$

Insertion of Equation 3.5 into the Young Equation 3.4 leads to:

Equation 3.6 $$\gamma_S - \pi = \gamma_{SL} + \gamma_L \cos\theta$$

Theoretically, the spreading pressure can be calculated from adsorption isotherm data if it is known to what extent which type of molecules are adsorbed on a surface [5]. In practice, however, the presence of moisture will change on a day to day basis, depending upon temperature and relative humidity, so it is very difficult to predict the influence of spreading pressure in a quantitative manner. Spreading pressure can, however, play a significant role. In the case of carbon black pigments, free surface energies which were found to lie in the order between 43 and 115 mN/m were lowered to values between 27 and 53 mN/m when n-hexane was adsorbed onto them [6]. For plate glass, free surface energies of 30 mN/m were measured under moisture saturated conditions, whereas at a ratio of water partial pressure to the partial pressure at saturation of 10^{-5}, the free surface energy was found to be about 75 mN/m [7]. The same paper reports that α-Al_2O_3 (sapphire) has a free surface energy of 1050 mN/m at 1500 °C and 10^{-5} Torr ($1.3332 \cdot 10^{-5}$ hPa) pressure, whereas at 0.6 % relative humidity and 20 °C the free surface energy falls to a value of 45 mN/m. When the relative humidity was further increased to 95 %, the free surface energy dropped down to 36 mN/m at room temperature. This shows that in the case of inorganic pigments, free surface energies that might have been measured have little to do with the actual solids, but reflect the conditions that are determined by the state of adsorption of small molecules which are mainly moisture or, otherwise, for example solvent molecules or other liquids used for measuring contact angles. In the case of polymers, the spreading pressure does not play such a great role because polymers have smaller surface energies to start with. Typical values lie somewhere between 15 mN/m (polytetrafluoroethylene) and 42 mN/m (polyethyleneterephthalate) [8], as shown in Table 3.1.

Table 3.1: Surface energies of some polymers

Polymer	Free surface energy [mN/m]
Polytetrafluorethylene	18.5
Polyvinylfluoride	30.3
Polyethylene	33.1
Polypropylene	31.2
Polystyrene	47.0
Polyvinylchloride	41.5
Poylamide 66	43.2
Polyethylene terephthalate	37.4
Polycarbonate	38.8
Polyacrylnitrile	44.0

In general, the free surface energy of a low-energetic surface is not lowered by the adsorption of substances with higher surface tensions.

If the Young equation is to be utilized at all, then it is necessary to measure contact angles. In the case of flat solid surfaces, a drop of liquid is carefully deposited with the aid of a syringe and a needle. The drop that is formed is often viewed with a special microscope that allows the drops to be observed from the side and that has the option to apply a tangent to the base point of the droplet. Optionally, the drops may be photographed so that contact angles can then be determined separately using image analysis tools (viz. Figure 3.3). It is often of interest to determine not only the contact angle of the proceeding droplet by stepwise addition of more and more liquid to the droplet, but also the contact angle of a receding droplet. This is done by carefully extracting liquid from the droplet through the needle. Contact angles of receding droplets are lower than the corresponding proceeding droplets since there are already adsorbed molecules at the surface so that the interfacial tension does not have to be overcome when the droplets cover less area of the substrate. The difference between the contact angle of a proceeding droplet to that of a receding droplet is called "hysteresis". A more detailed discussion of these phenomena is found in the literature [9].

3.3.1 Critical surface tension according to Zisman

It is often of interest to know the highest surface tension that a liquid may have to still wet a surface completely, that is to have a contact angle of zero. Zisman [10] suggested to determine this "critical surface tension" by plotting the cosine of the contact angles that droplets of liquids of homologous rows form on a surface against their surface tensions. Homologous rows are chemically similar molecules that nevertheless have different surface tensions. Examples are alkanes of different molecular weight or the row of alcohols. Frequently, this plot yields a straight line or, in any case, a curve that can be extrapolated to the point where the contact angle is zero ($\cos\Theta = 1$). The critical surface tension that corresponds to this point can be read from the abscissa (see Figure 3.5). The critical surface tension is somewhat dependent upon the type of homologous series used. In

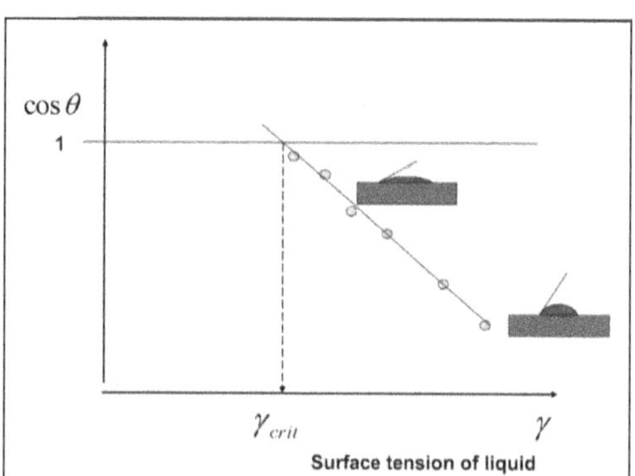

Figure 3.5: Determination of critical surface tension according to Zisman

materials engineering, test inks of different surface tensions are used, for example, to determine the critical surface tension of car paints in the spray booths within the assembly line. They normally consist of aqueous, coloured inks in which the surface tensions are adjusted by the addition of suitable organic liquids that lower the surface tension of water without being all too volatile.

At the critical surface tension Equation 3.4 turns to:

Equation 3.7 $$\gamma_S = \gamma_{SL} + \gamma_L$$

Complete wetting always takes place if $\gamma_S \geq \gamma_{SL} + \gamma_L$. The difference:

Equation 3.8 $$\gamma_S - (\gamma_{SL} + \gamma_L)$$

is called the spreading coefficient [11]. The speed of wetting depends not only on this difference but also to a high degree on the viscosity and on the roughness of the surface that is to be wetted. Rough, structured surfaces enhance the speed of wetting due to capillary effects.

As already stated, measuring the surface tension γ_L of a liquid is simple and, at least in the case of plane surfaces, this is also true for the measurement of contact angles. The critical problem of the Young Equation 3.4 is however that it embodies two unknown quantities, namely γ_S and γ_{SL}. This circumstance prevented the determination of free surface energies of solids for a long time. To resolve this shortcoming, a number of attempts were made to estimate the interfacial tension γ_{SL} from known quantities.

3.3.2 Approach of Good und Girifalco

Good und *Girifalco* [12] proposed to express the interfacial tension γ_{SL} as a function of the surface tensions of both the liquids and the solids and an interaction parameter Φ.

Equation 3.9 $$\gamma_{SL} = \gamma_S + \gamma_L - 2 \cdot \Phi\,(\gamma_S \cdot \gamma_L)^{1/2}$$

The interaction parameter is calculated from the molar volumes V_S and V_L of the solids and the liquids involved.

Equation 3.10 $$\Phi = \frac{4V_S^{1/3} \cdot V_L^{1/3}}{V_S^{1/3} + V_L^{1/3}}$$

By inserting Equation 3.9 into the Young equation 3.4 followed by rearranging, an expression is found that enables the determination of the free surface energy of a solid from the surface tension of the liquid and its contact angle on the solid.

Equation 3.11 $$\gamma_S = \left(\frac{\sqrt{\gamma_L}\,(1 + \cos\Theta)}{2\,\Phi}\right)^2$$

3.3.3 Approach of Fowkes

A principally new approach was introduced by *Fowkes* [13]. His assumption was that free surface energies are put together in an additive manner from contributions arising from dispersive and polar interactions. Dispersive interactions are formed by all substances, whereas polar interactions are inherent only for substances that either have permanent dipole moments or that are able to form hydrogen bonds (see Chapter 1).

Equation 3.12 $$\gamma = \gamma^d + \gamma^p$$

By breaking the surface tension down into a polar (index p) and into a dispersive (index d) contribution, the polar part of the surface tension of a substance for example can be determined from the total surface tension and its dispersive part if the latter can be accessed independently. Such an approach makes a lot of sense as can be shown by the following observation: If polystyrene, a nonpolar polymer, is oxidized by irradiation with UV-light in the presence of oxygen, then the contact angle of 1-bromo naphthalene, a nonpolar liquid, is not altered, whereas droplets of water show continuously smaller contact angles with increasing exposure times [14]. The only explanation for this is that polar groups are formed on the surface of the polymer in the course of the experiment and that these polar groups can undergo additional interaction only with other polar groups, yet not with nonpolar molecules[6]. Consequently, the contact angles of nonpolar liquids result solely from dispersive interactions, irrespective of if they are brought onto polar or nonpolar surfaces.

Fowkes predominantly examined nonpolar solid surfaces and the interactions between mercury and nonpolar liquids. Since mercury is a liquid at room temperature, its surface tension can be measured just like any other liquid. On the other hand, mercury is immiscible with nonpolar, organic liquids so that contact angles between the two can be measured. Just like any other substance, mercury undergoes dispersive interactions. Furthermore, it is in principle capable of forming metallic bonds, yet naturally not to nonpolar liquids. Fowkes started out with the following perception of the interface mercury/organic liquid (see Figure 3.6).

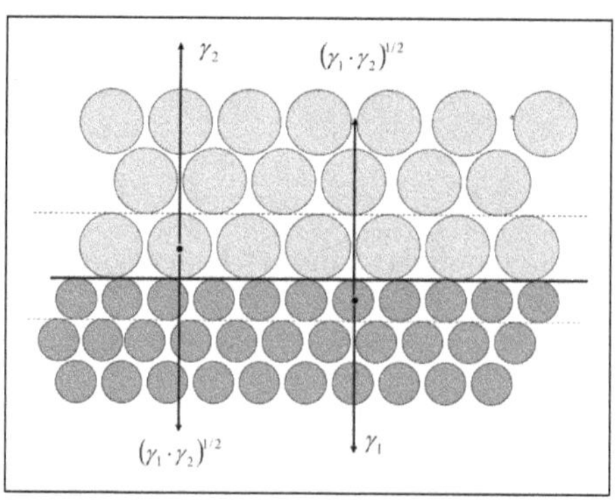

Figure 3.6: Interfacial tension according to Fowkes

[6] The oxidation of nonpolar substrate surfaces is technically used to improve the adhesion of printing inks and other coatings onto them.

Let the spheres in Figure 3.6 depict mercury atoms and alkane molecules (for example n-hexane). Since they don't mix, they form an interface. The interfacial tension is determined by the forces that act on the two types of molecules at the interface. If the mercury is indexed as 1 and the alkane as 2, then the mercury atoms at the interface are drawn towards the mercury by their own surface tension γ_1. On the other hand, they also undergo dispersive interactions with the molecules of hydrocarbons on the adjacent side of the interface. Fowkes [13] just like Good and Girifalco [12] reasoned that this component of interaction could be expressed by the geometric mean of the dispersive parts of the surface tensions. Since both contributions act in opposite directions, they must be subtracted from one another to find the net tension. The contribution of the mercury atoms at the interface to the interfacial tension is therefore $\gamma_1 - \sqrt{\gamma^d_1 \cdot \gamma^d_2}$. Analogously, the contribution of the alkane molecules is: $\gamma_2 - \sqrt{\gamma^d_1 \cdot \gamma^d_2}$. The resulting interfacial tension is the sum of both contributions.

Equation 3.13 $$\gamma_{12} = \gamma_1 + \gamma_2 - 2\sqrt{\gamma^d_1 \cdot \gamma^d_2}$$

If the polarizibilities α, the ionization energies I, the mean separation distances of the molecules r_{11} as well as the number of molecules per volume unit N is known, then, according to Fowkes [13] the dispersive part of the surface tension can be estimated by using Equation 3.14.

Equation 3.14 $$\gamma^d = \frac{\pi \cdot N^2 \cdot \alpha^2 \cdot I}{8 r_1^2}$$

In the case of water, for example (α = 1.48 Å^3; I = 12.6 V; $N = 3.34 \cdot 10^{22}/\text{cm}^3$; r_{11} = 2.76 Å), the dispersive part of the surface tension of 25.4 mJ/m² is calculated. With this value for the dispersive part and with the total surface tension, the polar portion of the surface tension may be easily calculated from Equation 3.12. Table 3.2 shows accordant values for a number of other liquids. The quotes may be somewhat different from source to source, depending upon the method employed to generate the data.

Table 3.2: Surface tensions of liquids

Liquid	γ	γ^d	γ^p
Water	72.8	21.8	51.0
Glycerine	63.4	37.0	26.4
Dijodomethane	50.8	44.1	6.7
Ethylenglycole	47.7	30.9	16.8
1-Bromonaphthaline	44.6	44.6	0
1-Octanole	27.6	21.3	6.3
n-Dodekane	25.4	25.4	0

Insertion of Equation 3.13 into the Young Equation 3.4 followed by rearrangement yields:

Equation 3.15

$$\cos \Theta = -1 + 2\sqrt{\gamma^d_S}\left(\frac{\sqrt{\gamma^d_L}}{\gamma_L}\right)$$

Equation 3.15 represents a straight line when the cosine of the contact angle is plotted against the quotient of $\sqrt{\gamma^d_L}/\gamma_L$. The dispersive part of the free surface energy of the solid is found from the slope. In theory, a single contact angle measurement is sufficient to calculate γ^d_S. But, with regard to the error inherently involved in measuring contact angles, a good recommendation is to make measurements with different liquids and then compare the γ^d_S data with each other.

Equation 3.13 may also be used to determine the dispersive contribution of the surface tension of polar liquids if both the interfacial tension and the dispersive part of the surface tension of one of the two partners are known. Interfacial tensions between immiscible liquids are easily found using the "spinning drop" method. In practice it may however present a problem to find pairs of liquids that are really completely immiscible.

3.3.4 Approach of Owens und Wendt

In case the substances forming an interface do not undergo solely dispersive interactions but polar interactions as well, then these contributions to the surface tensions have to be put into account also. Analogous to the reasoning and the steps that led to Equation 3.13 (see Figure 3.6 also), the interfacial tension may be expressed as:

Equation 3.16 $$\gamma_{12} = \gamma_1 + \gamma_2 - 2\sqrt{\gamma^d_1 \cdot \gamma^d_2} - 2\sqrt{\gamma^p_1 \cdot \gamma^p_2}$$

Inserting Equation 3.16 into the Young Equation 3.4 leads to

Equation 3.17 $$\gamma_L (1 + \cos \Theta) = 2(\sqrt{\gamma^d_S \cdot \gamma^d_L} + 2\sqrt{\gamma^p_S \cdot \gamma^p_L})$$

By measuring the contact angles of two different liquids whose dispersive and polar parts of the surface tension are known, the dispersive as well as the polar parts of the free surface energy of the solid can be determined [15]. For the sake of accuracy, one normally employs a graphic method for determining the free surface energy of a solid [16]. To achieve this, Equation 3.17 is divided by $2 \cdot \sqrt{\gamma^d_L}$. This once again yields a straight line.

Equation 3.18 $$\frac{\gamma_L (1 + \cos \Theta)}{2\sqrt{\gamma^d_L}} = \sqrt{\gamma^d_S} + \sqrt{\gamma^p_S} \cdot \frac{\sqrt{\gamma^p_L}}{\sqrt{\gamma^d_L}}$$

Equation 3.18 is known as the *Owens, Wendt, Rabel, Kälble (OWRK)* equation. When the contact angles of different liquids are determined and the left hand side of Equation 3.18 is plotted against the quotient between the square roots of the polar and dispersive parts of the surface tensions of the test liquids, then a line with an intersect with the y-axis at $\sqrt{\gamma^d_S}$ and the slope of $\sqrt{\gamma^p_S}$ results. Figure 3.7 shows such a graph that was found when measuring the free surface energy of a water based OEM automobile primer surfacer while using the test liquids water, glycerol, ethylene glycol, 1-octanol and n-dodecane. A free surface energy of

$29.5 \cdot 10^{-3}$ N/m was found with a dispersive contribution of $20.2 \cdot 10^{-3}$ N/m and a polar part of $9.3 \cdot 10^{-3}$ N/m. The tenacity of such measurements can be rated on the basis of how well the graphical representation of the experimental results follow a straight line. Care must be taken that the chosen test liquids do not swell the solid, since otherwise false data will be the result. Swelling can easily take place when polymeric substrates under investigation.

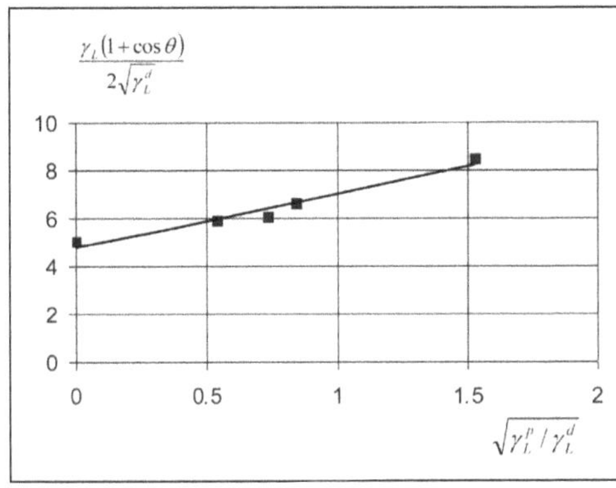

Figure 3.7: Measuring the dispersive and polar parts of the free surface energy of a water based OEM automobile primer (OWRK equation)

3.3.5 Approach of Wu

Wu [17] studied polymers that were able to undergo polar next to dispersive interactions. Unlike Owens and Wendt [15, 16], he chose the harmonic mean of the dispersive and polar contributions of the surface tensions to describe the interfacial tension. The harmonic mean between two numbers a and b is 2ab/(a+b). So it is the quotient between the square of the geometric mean $\sqrt{a \cdot b}$ and the arithmetic mean *(a+b)/2*. Following the same steps as for the derivation of Equation 3.13 and 3.16, the approach of Wu yields

Equation 3.19
$$\gamma_{SL} = \gamma_L + \gamma_S - \frac{4\,\gamma^d_L \cdot \gamma^d_S}{\gamma^d_L + \gamma^d_S} - \frac{4\,\gamma^p_L \cdot \gamma^p_S}{\gamma^p_L + \gamma^p_S}$$

Inserting Equation 3.19 into Equation 3.4 and rearranging leads to

Equation 3.20
$$\gamma_L\,(1 + \cos\Theta) = \frac{4\,\gamma^d_L \cdot \gamma^d_S}{\gamma^d_L + \gamma^d_S} + \frac{4\,\gamma^p_L \cdot \gamma^p_S}{\gamma^p_L + \gamma^p_S}$$

Unfortunately, it is not possible to change Equation 3.20 in a way that a graphical analysis can be carried out. Therefore, Wu used only two liquids, methylene iodide and water, to determine the dispersive and polar contributions of the free surface energies of some "polar" polymers. It is however possible to rearrange Equation 3.20 so that the fraction lines are abolished. This equation also connects the dispersive and polar parts of the free surface energy of the substrates with the corresponding values for the test liquids. The values for water and for methylene iodide are: $\gamma^d_{H20} = 22.1$ mN/m; $\gamma^p_{H20} = 50.7$ mN/m; $\gamma^d_{H20} = 44.1$ mN/m; $\gamma^d_{CH2J2} = 44.1$ mN/m; $\gamma^p_{CH2J2} = 6.7$ mN/m. For test liquids 1 (index 1) and 2 (index 2) the following equations are generated

Equation 3.21

$$(b_1 + c_1 - a_1) \cdot \gamma^d{}_S\gamma^p{}_S + c_1(b_1 - a_1) \cdot \gamma^d{}_S + b_1(c_1 - a_1) \cdot \gamma^p{}_S - a_1 \cdot b_1 \cdot c_1 = 0$$
$$(b_2 + c_2 - a_2) \cdot \gamma^d{}_S\gamma^p{}_S + c_2(b_2 - a_2) \cdot \gamma^d{}_S + b_2(c_2 - a_2) \cdot \gamma^p{}_S - a_2 \cdot b_2 \cdot c_2 = 0$$

in which

$$a_1 = \frac{1}{4}\gamma_1(1 + \cos\Theta_1) \quad b_1 = \gamma^d{}_1 \quad c_1 = \gamma^p{}_1 \quad \Theta_1 = \text{contact angle of liquid 1}$$

$$a_2 = \frac{1}{4}\gamma_2(1 + \cos\Theta_2) \quad b_2 = \gamma^d{}_2 \quad c_2 = \gamma^p{}_2 \quad \Theta_2 = \text{contact angle of liquid 2}$$

As, with the exception of $\gamma^d{}_S$ and $\gamma^p{}_S$, all variables are known, Equation 3.21 represents a system of two equations with two unknown variables. Since Equation 3.21 contains mixed terms as well as simple terms, it may not be solved by the subtraction method. So one must solve one of the two expressions either towards $\gamma^p{}_S$ or $\gamma^d{}_S$ and insert the result in to the other expression. This leads to a quadratic equation that is easily solved. As this is not explicitly described in the original literature [17, 18], the procedure is shown in appendix 1 of this book.

Table 3.3 compares the free surface energies of some polymers that were determined using the method of Wu and of Owens and Wendt [17], respectively, and some data of critical surface tensions using the concept of Zisman. All in all there is a good conformity between the free surface energy data according to Wu and to Owens and Wendt. However, the polar fractions of the surface energies are always lower when determined with the method of Owens and Wendt than when calculated according to Wu. Table 3.3 also includes some data that were gained from interfacial tension measurements between molten polyethylene and melts of other polymers. The method lying behind this is found in the literature [18]. These values follow the data gained from the Wu method significantly better than from those found according to Owens and Wendt. Still, there is a remarkable correlation between the critical surface tensions from Zisman plots and the free surface energies calculated both from the Wu as well as the Owens and Wendt method. Grave differences are only obvious in the cases of polystyrene and the polyvinyl-polymers where the critical surfaces tensions are lower than the free surface energies found using the other methods. The Zisman evaluation delivers only a measure for the total free surface energy and does not distinguish between polar and dispersive contributions. Furthermore, it is unknown as to what extent the polymers used are actually comparable since the data were put together from different sources [17].

3.3.6 Interfacial tension at complete wetting

There is a controversy on whether in the case of complete wetting, that is when the contact angle becomes zero, the interfacial tension also comes to zero or not. If

Table 3.3: Calculated and determined surface tensions of polymers in comparison

Polymer	contact angle at 20 °C		Calculated surface energies mN/m				Experimentally determined surface energies mN/m		
			Wu		Owens and Wendt		Extrapolated values from polymer melts		Critical surface tensions
	H_2O	CH_2J_2	γ	γ^d/γ	γ	γ^d/γ	γ	γ^d/γ	γ_c
Polyethylene	102	53	36.1	0.98	33.2	1.00	35.7	1.00	31
Polytetrafluorethylene	108	77	22.5	0.91	19.1	0.97	26.5		19
Polystyrene	91	35	42.6	0.90	42.0	0.98	40.7	0.83	33
Poly (methyl methacrylate)	80	41	41.2	0.75	40.2	0.89	41.1	0.72	40
Poly (n-butyl methacrylate)	91	52	34.6	0.82	33.3	0.94	31.2	0.84	32
Poly (chlorotrifluoroethylene)	90	64	30.1	0.72	27.5	0.87	31.1		31
Poly (vinylidene fluoride)	82	63	33.2	0.62	30.3	0.77			25
Poly (vinyl fluoride)	80	49	38.4	0.71	36.7	0.85			28
Poly (vinylidene chloride)	80	29	45.4	0.80	45.0	0.93			40
Poly (vinyl chloride)	87	36	41.9	0.85	41.5	0.96			39
Poly (ethylene terephthalate)	81	38	42.1	0.78	41.3	0.92			43
Poly (hexamethylene adipamide)	72	28	47.9	0.73	47.0	0.87			46

that were the case, then, according to the Young equation (Equation 3.4), the free surface energy would become identical to the surface tension of the liquid that just shows complete wetting, so $\gamma_S = \gamma_L$. *Neumann* and *Sell*[19] support the assumption that "the interfacial tension between a solid and a liquid approach zero in line with the contact angle". This would mean that the critical surface tension according to Zisman would be identical to the free surface energy of the solid. This would also enable the calculation of the interfacial tension γ_{SL} from this value and from the contact angle Θ and the surface tension γ_L of any liquid using Equation 3.4.

In contrast, Owens and Wendt[15] pointed out that this assumption is only feasible if both the test liquid and the solid are nonpolar. When a polar liquid is brought into contact with a nonpolar solid or, vice versa, a nonpolar liquid with a polar solid, then the critical surface tension is lower than the free surface energy of the solid. In other words: $\gamma_{SL} > 0$ when complete wetting occurs. Table 3.3 actually does demonstrate that critical surface tensions are all somewhat smaller than the free surface energies. However, this also holds true for nonpolar polymers such as polyethylene. As pointed out, Table 3.3 was put together from different sources, so that the comparability of the different polymers is not necessarily established to one hundred per cent. But, even if the discrepancies are systematic, the small differences do show that the determination of the critical surface tension as a measure for the free surface energy of solids is a practical and a reasonably meaningful method.

3.4 Wetting of pigments

The importance of the wetting of agglomerates for the dispersion process was already mentioned in Chapter 3.1. However, a reduction in agglomerate strength only takes place, if all of the void volume within the agglomerates is filled with liquid. Lesser wetting even lets agglomerates become tougher. We all know this phenomenon from the times when we were still playing in the sand-box. In order to make a decent cake of sand with a mould, the sand had to be somewhat wet. Neither completely dry sand, nor sand that is too wet is useful for this purpose because both a dry powder as well as slurry will flow under the influence of gravity. *Schubert*[20] studied the influence of moisture on the cohesiveness of powders by pressing powders with defined moisture contents into tablets followed by a tensile test. Figure 3.8 shows the strength of the tablets as a function of the degree of saturation of the void volume with moisture in a qualitative manner. Three different regions may be distinguished. In the first region A, the liquid does not form a continuous phase. The liquid molecules can best be referred to as adsorbed molecules of gaseous water. In this region the water acts like an adsorbed surface treatment component on a pigment. That is why the agglomerate strength declines with increasing surface coverage. (Schubert postulated an

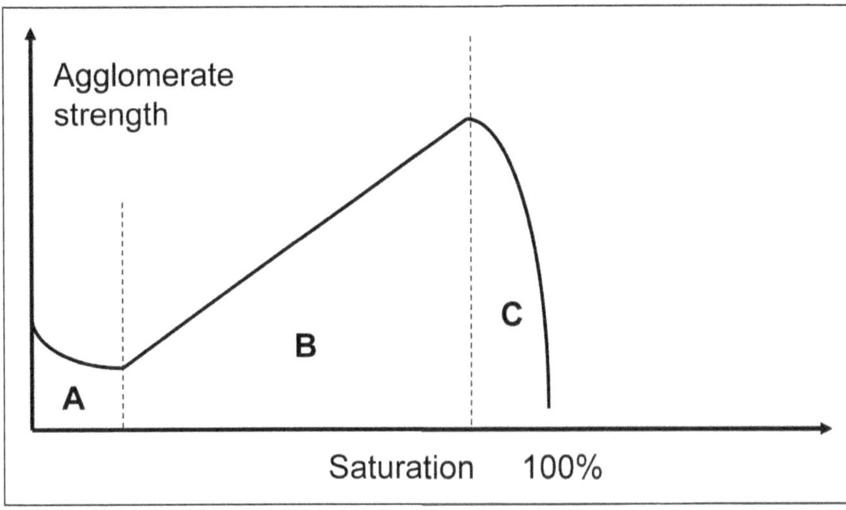

Figure 3.8: Three different regions of agglomerate strength with increasing liquid content
A: the liquid does not form a continuous phase
B: liquid bridges appear between the particles
C: the void volume within the agglomerates is completely filled with liquid

increase in strength in this region. This was questioned by *Winkler* [1] and was rebuted by *Vervoorn* [21] experimentally.) In the following region B, liquid bridges appear between the particles. These solidify the agglomerates because when the particles are separated, the surface area of the liquid bridges increase so that work has to be applied against their surface tension [22]. Pigment developers make use of this fact by rinsing wet pigment samples with, for example, methanol before drying them. Since methanol has a surface tension of 22.6 mN/m instead of 72 mN/m, like water, the dried pigment cake is then loosely packed and may be dispersed in a polymeric system for further evaluation without a prior milling step. Finally, in the region C, the void volume within the agglomerates is completely filled with liquid. The strength of the agglomerates then falls abruptly to a very low value.

When dry pigment agglomerates are brought into a liquid, then the liquid enters the void volumes of the agglomerates from all sides due to the capillary pressure. When this happens, the entrapped air within the agglomerates is compressed. The entry of the liquid into the agglomerates comes to a halt once the pressure of the trapped air is as large as the sum of the atmospheric pressure and the capillary pressure. If the agglomerates remain intact under this pressure, then the wetting process stops and further infiltration of liquid may only proceed to the extent as the entrapped air is dissolved in the liquid.

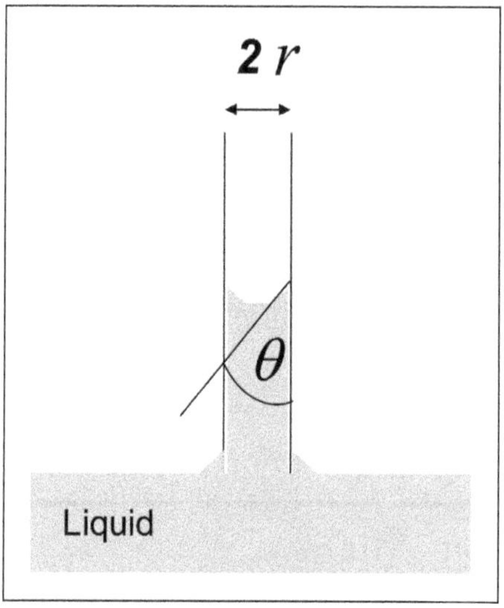

Figure 3.9: Liquid in a capillary

The capillary pressure P_K that forms in a round capillary (see Figure 3.9) of radius r is given by

Equation 3.22 $$P_K = \frac{2 \cdot \gamma_L \cdot \cos \Theta}{r}$$

Figure 3.10 shows the influence of the porosity and the contact angle of a liquid on a pigment surface on the entry of the liquid into agglomerates in a qualitative manner. Using the ideal gas law:

Equation 3.23 $$P_1 \cdot V_1 = P_2 \cdot V_2$$

The relative volume change V_1/V_2 may be calculated as the pressure is increased from a value of P_1 to $P_2 = P_1 + P_K$ and the relative diameter d_2/d_1 of the dry core of an agglomerate can be estimated from Equation 3.24.

Equation 3.24 $$\frac{d_2}{d_1} = \sqrt[3]{\frac{V_2}{V_1}}$$

In the derivation of Equation 3.34, it was assumed that the agglomerates show the same porosity at all places and also that the void volume of an agglomerate acts like a multitude of small capillaries. As for the surface tension, the value for water, namely 72 mN/m, was applied. Figure 3.10 demonstrates that wetting is facilitated especially when the capillary radii become small, that is with decreasing agglomerate porosities, since the dry cores of the agglomerates become smaller. It can also be seen that liquids showing complete wetting ($\cos \Theta = 1$) perform best, whereas wetting becomes less when the contact angles become greater. So, wetting behaviour and mechanical strength of agglomerates are oppositely affected by agglomerate porosity.

A loosely packed agglomerate with a large void volume has fewer contact points amongst the primary particles and, in most cases, less adhesive strength at the contact points. As explained in Chapter 2, loosely packed agglomerates come into being when the attractive forces between the particles are lower. Which of the two influences, wetting behaviour or mechanical strength of the agglomerates, have a larger influence, may be different from case to case. In general, even the strongest agglomerate will be dispersed, given that it is stressed strongly enough and any wetting procedure will eventually be quantitative as long as the contact angle is smaller than 90° and provided the liquid has enough time to do so.

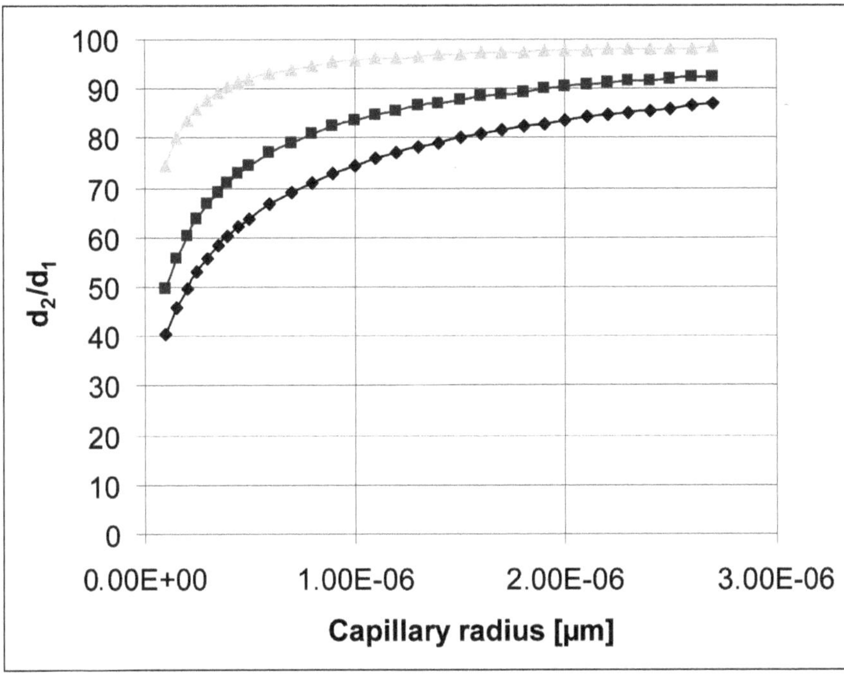

Figure 3.10: Influence of the porosity and the contact angle of a liquid on a pigment surface on the entry of the liquid into agglomerates.
a) cos θ = 1; θ = 0°
b) cos θ = 0.5 ; θ = 60°
c) cos θ = 0.1; θ = 84.3°
d_1= diameter of an agglomerate; d_2 = diameter of a dried agglomerate

3.4.1 Measuring the free surface energy of pigments

Until today, the measurement of free surface energies of pigments is more an academic than a practical exercise. Nevertheless, the findings that are gained are of general importance for the understanding of wetting processes.

Unlike plastics and other substrate materials, pigments mostly accrue as minute, in many cases sub-microscopic particles. One of the first methods for measuring the contact angles on powdery substances was developed by *Bartell* and *Osterhoff* [23]. The authors constructed a cylindrical measuring cell into which the powders under investigation can be pressed in a defined way. Both sides of the measuring cell are sealed with frits that let liquids and air pass, but otherwise keep the powder in its compressed form. One side of the measuring cell is brought into contact with the test liquid. Under these conditions, due to the capillary pressure, the liquid penetrates into the enclosed powder bed. The entry of the liquid into the powder is terminated by applying a back pressure onto the opposite side of the measuring cell. The pressure necessary to achieve that is identical to the capillary pressure. This may be first done with a number of liquids that have very low surface tensions, so that one may assume the contact angle to be zero. Then it is possible to determine r using Equation 3.22. Of course, under these circumstances, r is no longer the radius of a capillary but should be understood as a capillary factor that reflects the influence of the packing of the powder. In a next step, the capillary pressure P_K of a liquid forming a contact angle greater than zero can be determined and the corresponding contact angle can be calculated utilizing the capillary factor r that was formerly found. A precondition for meaningful measurements is naturally that the powder can be compressed to a defined porosity in a reproducible manner.

A similar method was developed by *Washburn* [24]. Washburn combined the capillary pressure Equation 3.22 with the Hagen-Poiseuille equation for the rate of entry of a liquid into a capillary (Equation 3.25). According to the Hagen-Poisseuille equation, the speed $\Delta l/\Delta t$ with which a liquid is sucked into a capillary is proportional to the pressure difference ΔP as well as the square of the capillary radius r and inversely proportional to its length l and to the viscosity η of the liquid.

Equation 3.25
$$\frac{\Delta l}{\Delta t} = \frac{r^2 \cdot \Delta P}{8 \cdot \eta\, l}$$

Inserting Equation 3.22 into Equation 3.25 (with $\Delta P = P_K$) yields a further expression for the rate of entry of a liquid into a capillary. For n capillaries one may write:

Equation 3.26
$$\frac{\Delta l}{\Delta t} = \frac{n \cdot r \cdot \gamma_L \cdot \cos \Theta}{4 \cdot \eta \cdot l}$$

The volume of liquid V_L that rises into a capillary is given by:

Equation 3.27
$$V_L = 2\pi \cdot r^2 \cdot \Delta l$$

Expanding Equation 3.25 by the factor $2\pi \cdot r^2$ and inserting $V_L/(2n\pi \cdot r^2)$ in the place of l (from Equation 3.27) followed by a transformation leads to

Equation 3.28
$$V^2_L = \frac{\pi^2 n^2 r^5 \gamma_L \cos \Theta}{2\eta} \cdot t$$

It is easier and more accurate to measure an increase in weight with a weight balance rather than a volume increase. Thus, replacing the liquid volume V_L by the quotient between the mass m_L and the density ρ_L of the liquid and then forming the square root of the expression leads to the Washburn equation:

Equation 3.29
$$m = \sqrt{\frac{n^2 r^2 \pi^2}{2}} \cdot \rho \cdot \sqrt{\frac{\gamma_L}{\eta}} \cdot \sqrt{\cos \Theta} \cdot \sqrt{t}$$

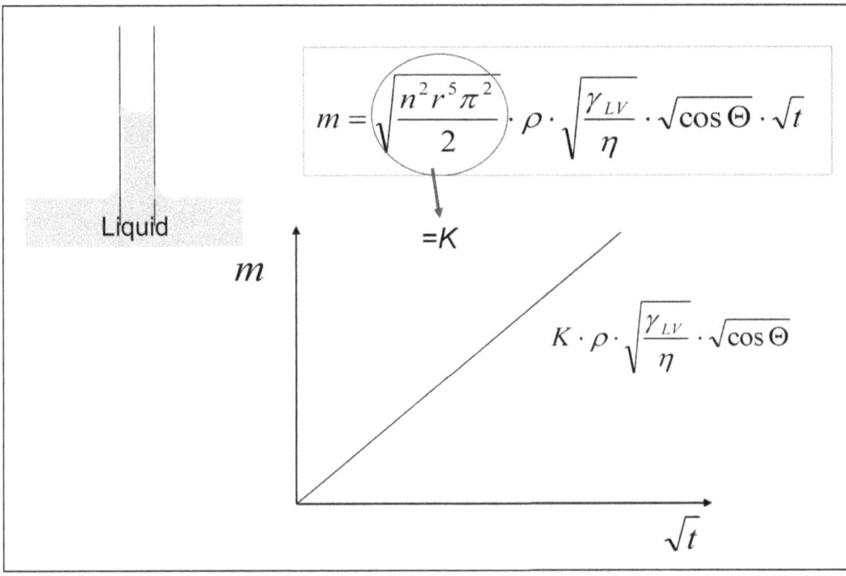

Figure 3.11: Contact angle of pigment surfaces measured with the Washburn method

The first expression on the right hand side of Equation 3.29 depicts the capillary term K which, just like when using a Bartell cell, can be found by first measuring the rate of entry of a number of liquids that show complete wetting, so that $\cos \Theta = 1$. For the following measurements using non-spreading liquids, this value of the capillary term is used for calculations. According to Equation 3.29, a straight line is found when plotting the mass uptake of the liquid into the compacted powder against the square root of the time. The contact angle may be found from the slope of the line, given that the capillary term as well as the liquid properties: surface tension, viscosity, density are known. This is shown schematically in Figure 3.11.

Just like in the Bartell method, a reproducible pigment packing procedure is the most crucial requirement when it comes to generating meaningful data using the Washburn procedure. Although this presupposition is not easily met in all cases, the Washburn method has become accepted and instruments for carrying out measurements are commercially available.

Table 3.4 lists free surface energies of some fillers that were gained by critical surface tension measurements following the Zisman method. The contact angles for this were determined using a Washburn cell[7] [25].

[7] Compare Table 3.3

Table 3.4: Free surface energies of some fillers

Filler	Free surface energy [mN/m]
Calcium carbonate (Millicarb)	39
Amorphous silica (Silbond FW 12)	52
Quartz with rounded edges (Novacite 200)	65
Splintery, ground Quartz (Silbond W12)	49
Aluminum hydroxide (treated with epoxisilane)	53

A further method for determining free surface energies of pigments is based upon the direct measurement of contact angles on compact powder tablets. *Wu* and *Brzozowski* [26] compressed organic pigments to tablets using pressures between 345 and 965 bar. Under these conditions, tablets with mirror like surfaces were produced. Utilizing the algorithms of Wu (Equation 3.20), the dispersive and polar fractions of the surface energies were determined. Once again, water and methylene iodide were used as test liquids. Table 3.5 shows the results for the different organic pigments. Notably, the dispersive contributions of the free surface energies prevail while total free surface energies between 40 and 65 mN/m were found. These values harmonize well with the free surface energies of polymers (see Tables 3.1 and 3.3)

Using high pressures, inorganic pigments like titanium dioxide may also be compressed to specular tablets. However, they are not excessively mechanically stable and tend to be soaked by liquids more or less quickly. Contrarily, in the case of

Table 3.5: Dispersive and polar fractions of the free surface energies of different organic pigments (Source: [26])

Pigment	γ^d	γ^p	γ	γ^p/γ
Indanthrone	33.2	30.0	63.2	0.48
Thioindigo red	35.1	16.3	51.4	0.32
Isoindolinone	32.2	15.0	47.2	0.32
Gamma Quinacridone	35.7	13.4	49.1	0.27
Toluidene red	39.7	13.3	53.0	0.25
Phthalocyanine (metal free)	40.1	12.7	52.8	0.24
Chlorinated Cu-phthalocyanine	35.8	6.2	42.0	0.15
Cu-phthalocyanine	40.0	6.9	46.9	0.15

organic pigments, it might be that the pigment particles coalesce by cold fusion under pressure, thus forming solid surfaces that are appropriate for such measurements.

A further method for the determination of free surface energies of powdery substances was proposed by *Kossen* and *Heertjes* [27]. It is known as the *"h-ε method"*. Once again, the powders under investigation are compressed to tablets. Yet, instead of applying a droplet, the liquid is continuously fed to the surface of the tablet until a "lake" is formed whose height h remains constant upon further addition of liquid. The contact angle is calculated from the height h of the liquid meniscus, the porosity ε of the tablet, the surface tension γ_L and the density ρ_L of the liquid and from the gravitational acceleration g using Equation 3.30

Equation 3.30
$$\cos \Theta = 1 - \sqrt{\frac{Bh^2}{3(1-\varepsilon)(1-Bh^2/2)}}$$

In which B is given by:

Equation 3.31
$$B = \frac{\rho_L g}{2\gamma_L}$$

Contact angles of various alcohols and aniline on powders of sodium chloride and water on compressed talc powders correlated with those found in direct measurements on plane surfaces of these substrates [28].

In the case of pharmaceutical powders, *Buckton* [29] found rapidly decreasing contact angle values with increasing pressures used for compacting. He ascribed this to the generation of fresh surface areas due to breakage of the powders. *Broeckel* and *Loeffler* [30] made a similar observation. They therefore proposed to glue he particles onto glass plates and take direct measurements on these specimens. This method bears the additional difficulty that the porosities of these layers of immobilized particles have to be known. They must be determined separately using image analysis techniques. The resistance of the layer of glue towards the test liquids constitutes a further constraint of the method.

Lerk und *Lagas* [31] avoided the problems of a possible change in the surface area of pharmaceutical powders during compression by incorporation of the powders into matrices that were formed into porous plates by applying only low pressures. Before the application of the test liquid (water) the tablets were sprayed with water until they were completely wetted. Then, the contact angle was measured using the method of *Kossen* and *Heertjes*. In their paper, the authors were able to correlate the release of the drugs in the body with their wetting behaviour, so one might expect this method to be applicable for pigment testing as well, provided that the tablets are at least stabile enough to avoid the particles to change places merely by the interaction with the test liquids[8].

[8] For example, in the case of titanium dioxide pigments that were surface treated with silicone oils, the author found the tablets to "explode" with a loud bang when brought into contact with water.

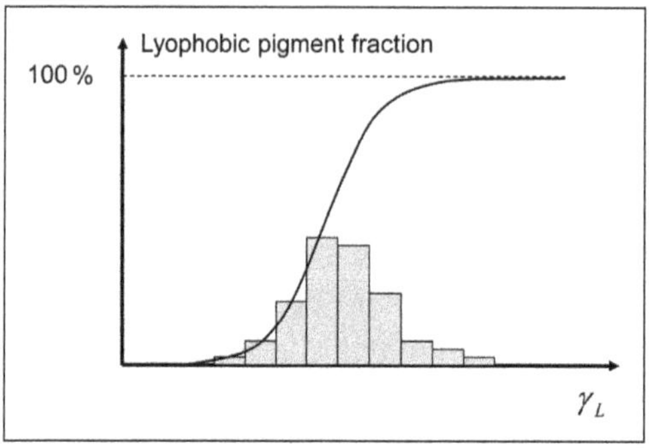

Figure 3.12: Determination of the free surface energy distribution of pigments by the immersion method

An immersion method for determining the surface energy distribution of carbon dust was described by *Fuerstenau* and *Williams* [32]. Mixtures of water and methanol were filled into a separating funnel with an upper diameter of 75 mm and an amount of carbon dust that just formed a mono layer of powder on the surface (0.06 to 0.3 g, depending on the sample) was added. Under these circumstances, particles having a contact angle of zero degrees are wetted and sink into the liquid, whereas those which form a contact angle greater than zero degrees remain afloat. By varying the composition of the aqueous alcohol solutions, surface tensions between 22.5 mN/m and 72.8 mN/m may be adjusted. In intervals of only a few mN/m of liquid surface tension, the floating fraction of pigment can be separated from the wetted portion. A plot of the amount of buoyant (hydrophobic) fraction as a function of the surface tension of the liquid mixture yields an S-shaped cumulative sum curve as shown schematically in Figure 3.12. Assuming the critical surface tension to be identical to the free surface energy of the particles, the frequency distribution of the free surface energy of the powder may be accessed by mathematical differentiation of this curve (see Figure 3.12). Of course, this method is limited to powders whose free surface energies lie within the boundaries of surface tensions covered by the liquid mixtures. But, fortunately, this does not present a serious limitation in practice. Although the use of liquid mixtures may give rise to questions due to the possibility of preferential adsorption of one of the components[9], the method nevertheless delivers useful indications as to the wetting behaviour of powders and has the advantage of being comparatively simple with no large expenses for test equipment being necessary.

A further method for determining the free surface energy of powdery substances is inverse gas chromatography [33–36]. For this, the powder under investigation is brought into a wide bore gas chromatography column. The dispersive part of the surface energy is gained from the retention of homologous rows of alkanes. The values of surface energy found tend to be very high, though. For carbon blacks, surface energies of up to 500 mN/m were found using this method [33]. This may be attributed to the fact that under the usual conditions of gas chromatography, namely temperatures in the range between 175 and 250 °C and inert gas flow, the

[9] When preferential adsorption takes place, the interfacial tension is altered.

surfaces are free of adsorbed molecules that may reduce the free surface energies (see Equation 3.5, "spreading pressure"). On the other hand, the free surface energy of particles is a distributed property. The energy will be higher at edges and rims of the pigment crystals than at flat surface parts. Since the adsorption occurs predominantly at locations with higher surface energies, these sites will contribute prevailingly to retention. This assumption is confirmed by the observation *Winkler*[37] made when studying commercially available iron oxide and titanium dioxide pigments in inverse chromatography: The retention time increased consistently with decreasing loadings of the analyses. He explained this with the presence of surface sites with different free surface energies.

Summing up, it can be ascertained that the free surface energy of solids determine their wetting behaviour. Yet, the determination of free surface energies of pigments is not particularly simple. Still, there are a number of methods available to characterize pigments and to compare them amongst each other. It must be kept in mind that the free surface energies can change within a wide range depending upon the relative humidity and possible exposition towards solvent vapours. Organic surface treatments that are selectively used to impart hydrophilic or organophilic properties onto pigments play a role in this context. In many cases, these additives disconnect from the pigments when these are put into formulations so that the interactions between the liquid phases and the pigments may be different after adsorption/desorption equilibria have been reached compared with the initial wetting step.

3.4.2 Kinetics and thermodynamics of pigment wetting

The Washburn equation (Equation 3.29) demonstrates the need to distinguish between the kinetics and the thermodynamics of wetting. In low viscous systems the liquid molecules are fast enough to enter the voids of the agglomerates quickly; given that the thermodynamic preconditions are opportune (that is high free surface energy of the pigments, low surface tension of the liquid, low interfacial tension between pigment and liquid and – as a result of this – a low contact angle). For that reason, one will always strive to let the wetting take place in a low viscosity medium. Examples for such low viscosity systems are coatings formulations. Since they are meant to spread on substrates, they tend to have low surface tensions anyway. Furthermore, wetting of agglomerates can be caused by solvent molecules alone, without dissolved polymer molecules. Especially solvent based paints exhibit low surface tensions. Aqueous systems are formulated with additives which reduce the surface tension.

Highly viscous systems are encountered in polymer melts, for example, in the production of powder coatings or thermo-softening plastics. Usually, the pigments are less well dispersed in these systems. The scattering power of titanium dioxide pigments in powder coatings, for example, is far worse than in solvent based coatings, which certainly owes to the kinetics of wetting, since polymer melts normally have surface tensions even lower than the critical surface tensions of the solidified polymers. Typical values lay between 30 and 50 mN/m, depending upon the type of polymer and the temperature (Tables 3.1 and 3.2). Liquid paints mostly have surface tensions between 25 and 35 mN/m. So, these systems would be expected to wet oxidic pigment surfaces completely as long as these were not altered unfavourably in terms of wetting by organic additives. Examples of oxidic pigments are titanium dioxide, iron oxide, chromium oxide, zink oxide, talc, mica etc. Yet, adsorbed silicone oils, for example, lower the surface energies of the pigments to such an extent that they are wetted very poorly, if at all. In the case or organic pigments, Table 3.5 shows that some are wetted better and some worse. It is customary to use "wetting agents" that reduce the surface tension of the liquid, yet probably even have a larger effect by lowering the interfacial tension between the pigment and the liquid phase.

3.4.3 Wetting volume

When a heap of pigment powder is brought into contact with a liquid, the powder starts to absorb the liquid. In doing so, the powder bed collapses more or less. The uptake of liquid happens "voluntarily", that is without any further assistance, to a certain, characteristic amount and ends when that point is reached. *Von*

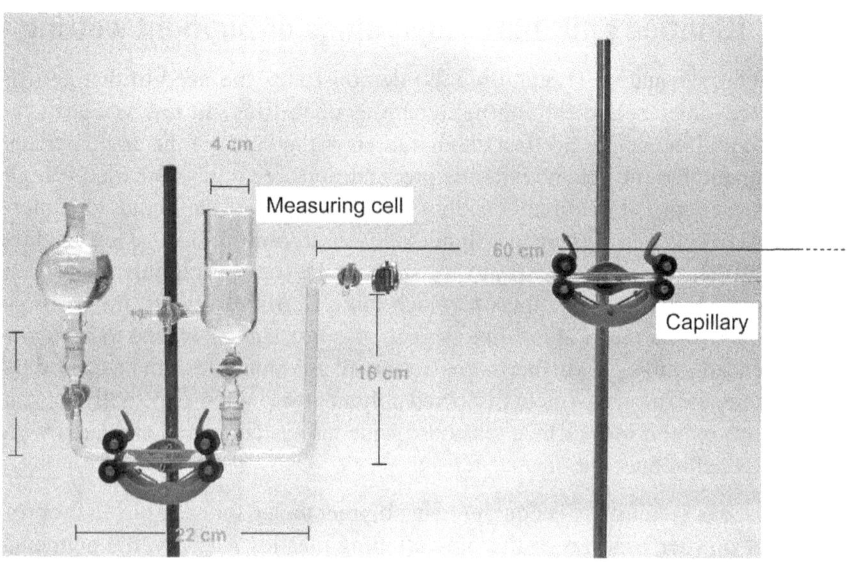

Figure 3.13: Apparatus for determining the wetting volume

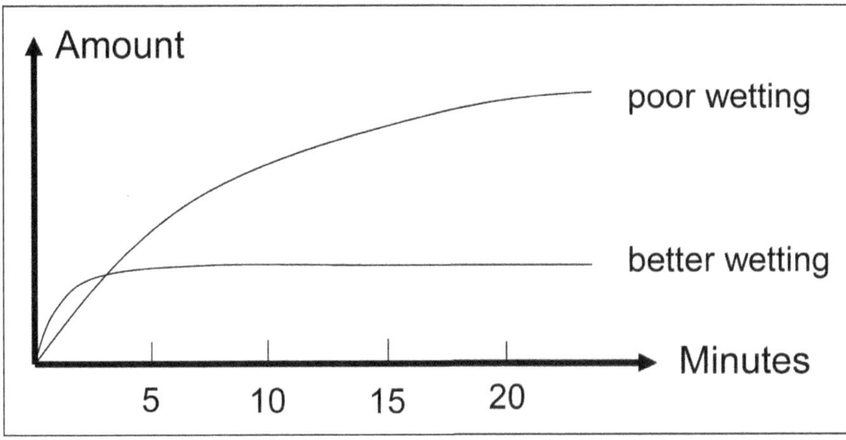

Figure 3.14: Determination of the wetting volumes for one liquid with better and one with poorer wetting behaviour

Stackelberg und *Frangen* [38] investigated this phenomenon and used the quotient between the absorbed volume of liquid and the amount of pigment powder used to describe the behaviour of the pigments. The so called "wetting volume" *WV* has the physical dimension *ml/g*.

Figure 3.13 shows an apparatus for determining the wetting volume. When carrying out a measurement, the pigment is placed on to the glass frit by a powder funnel. The glass frit itself is in contact with the test liquid from below (see Figure 3.13). By opening the valve below the frit, the liquid is able to follow, so that with the stop cock to the measuring capillary being open and the one to the reservoir closed, pigment wetting may take place. The uptake of liquid can be monitored at the capillary, for example, by placing a millimetre scale behind it. Figure 3.13 indicates the dimensions of the different components of the apparatus to enable a better impression.

Experiments show that different liquids need different amounts of time to penetrate into the pigment powders and also that different amounts of liquid are needed for complete wetting (Figure 3.14).

The steps that take place during wetting may be analyzed as follows: Driven by the often cited capillary action of the voids within the powders, liquid starts to penetrate into the powder bed. Unlike in the case of agglomerates which are thrown into a liquid, the entrapped air is able to escape in the upward direction, so the liquid can penetrate in an unhindered manner from below. Due to its surface tension, the advancing liquid front tries to minimize its interface towards the air. That is why the liquid, acting like a semi-liquid bridge, pulls the pigment particles together. The wetting volume therefore becomes less with increasing liquid surface tension γ_L. The contracting effect of the liquid also becomes more pronounced, the lower the contact angle Θ between the pigment and the liquid is. Since the wetting volume is by definition based on the mass of pigment used, the wetting volume necessarily has to fall with increasing pigment density ρ_P [10]. A more precise analysis [39, 40] led to the conclusion that the wetting volume is a function which, in a first approach, depends upon the surface tension of the liquid, the contact angle and the density of the pigments or fillers in a way shown in Equation 3.32.

Equation 3.32
$$BV = f\left(\frac{1}{\gamma^2_L};\frac{1}{cos^2\,\Theta};\frac{1}{\rho_P};P\right)$$

P in Equation 3.32 depends upon the particle size of the powder under investigation, so it differs from pigment to pigment. Experiments show that smaller particles tend to larger wetting volumes, probably because of the higher surface area that has to be wetted. The first layers of liquid are immobilized by adsorption and cannot contribute to the flow of the wetted powders.

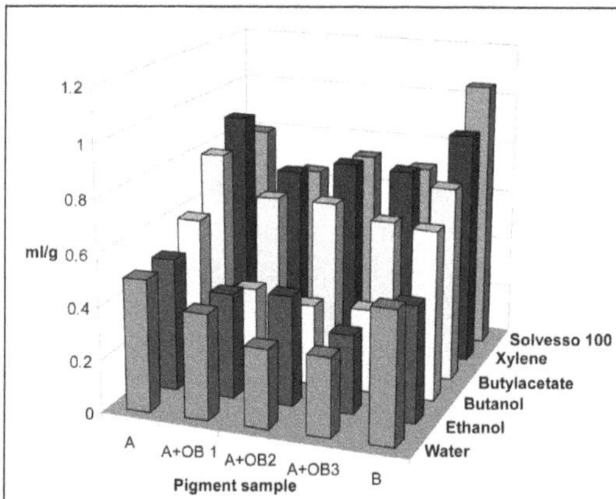

A: Pigment A without organic surface treatment
A+OB1: Pigment A plus 0.2 weight% trimethylolpropane + 0.3 weight% polyethylenglycole, MW 200
A+OB2: Pigment A plus 0.5 weight% ethoxylated trimethylolpropane, 90 to 100 % ethoxylation
A+OB3: Pigment A plus 0.5 weight% trimethylolpropane
B: Pigment B without organic surface treatment

Figure 3.15: Examples of wetting volumes for two titanium dioxide pigments with different inorganic surface treatments

In a technical sense, a liquid shows better wetting if it penetrates the powder bed faster and if it needs less liquid to wet the powder completely. The logic behind this is straight forward in terms of the speed of wetting. Considering the amount of liquid necessary, it is clear that more pigment can be introduced into liquids with low wetting volumes. Therefore, a mill base formulated with such a liquid will enable a higher pigment loading and will be more economical in terms of machine utilization.

Measurements of wetting volumes are very reproducible, easy to perform and

[10] This could be avoided by relating the wetting volume to the volume of pigment used. Then the physical dimension would be ml liquid/ml pigment.

give significant results. Thus, for example, the influence of inorganic and organic surface treatments on the wetting behaviour of titanium dioxide pigments can be determined flawlessly (Figure 3.15). By using liquids like water, alcohols, esters, aliphatic and aromatic hydrocarbons, a colloidal "finger print" of a pigment can be established, whereby similarities as well as differences of pigments can be accessed. In Figure 3.15, the titanium dioxide pigment A had aluminium oxide in the topmost layer of its inorganic surface treatment, whereas the pigment B had been treated with some silica as well. Figure 3.15 shows that the pigments have similar gradation in terms of wetting by the different liquids. The organic surface treatments used were typical for titanium dioxide pigments (see legend to Figure 3.15). As can be seen, they all improve the wetting of the pigment sample A.

Wetting volume measurements can also be used to determine the influence of additives on the wetting of liquids. Furthermore, it is possible to use resin solutions directly, provided their viscosity is not too high to suppress their penetration through the frits. It was shown [40] that the wetted pigment powders had less resistance towards the penetration of a test body (Penetrometer test) when the wetting volume was low and that the mill bases had improved flow. In the case of inorganic white pigments (rutile, anatase, zinc sulfide, zinc oxide), the wetting volume for toluene was lowered considerably when 0.1 mol of polar molecules (for example amines) were added. The reduction was more pronounced, the smaller the polar molecules were, i.e., the lower their space requirement on the pigments were.

3.5 Mill base rheology and mill base optimization

The flow of mill bases is an extremely important feature when producing paints. Not only the dispersion of pigments, but also the handling, like pouring and pumping of the paint, right up to the cleaning of containers is grossly effected.

Because of their consistency from dissolved polymers as well as pigments, fillers and various additives, mill bases as well as the completed paints generally tend to exhibit complicated flow properties. In this chapter, the fundamental terms used to describe viscous flow are first presented before the flow behaviour of mill bases and paints are discussed.

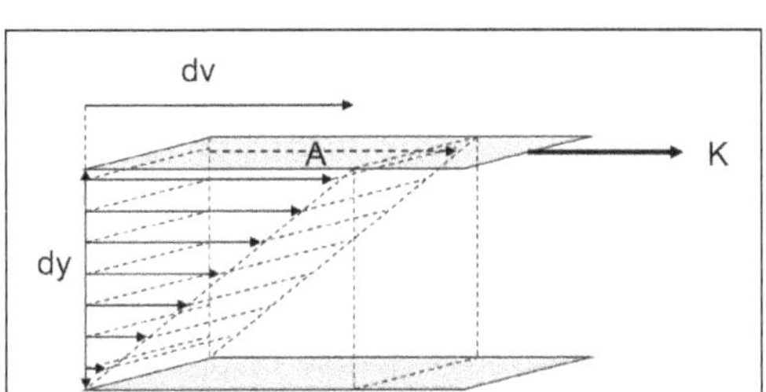

Figure 3.16: Model for viscosity

Figure 3.16 depicts the model behind the expression "viscosity". Two plates of area *A* are situated parallel to each other

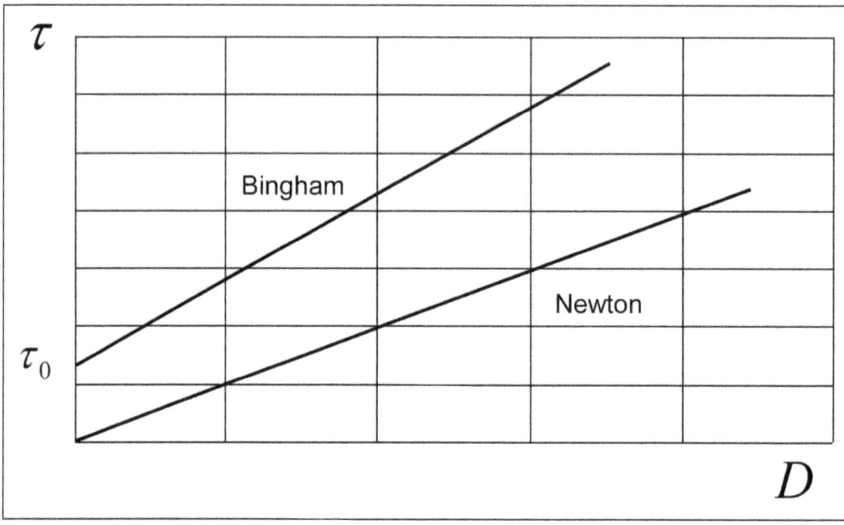

Figure 3.17: Newtonian and Bingham flow behaviour

at a distance *y*. In between the plates there is a liquid with normal, that is "Newtonian" flow behaviour, such as, for example, water. The top plate is pulled with force *K*, so that it moves relative to the bottom plate with velocity *v*, so that a velocity gradient *dv/dy* comes into being between the two plates. The force with which acts of the upper plate divided by its area *A* is proportional to the velocity gradient. The velocity gradient is called "shear rate" *D*. The quotient between the force and the area is termed "shear stress" τ. The proportionality constant is the dynamic viscosity η. It is a measure for the inner friction in the liquid, or better, for its resistance to flow.

Equation 3.33 $\tau = \eta \cdot D$ *or* $\eta = \frac{\tau}{D}$

The shear rate has the physical dimension $m/s \cdot m = 1/s$. The physical dimension of the shear stress is N/m^2. Accordingly, the dynamic viscosity has the dimension $N \cdot s/m^2 = Pa \cdot s$ (Pa = Pascal). The reciprocal value of the viscosity is called "fluidity".

When the viscosity is independent of the shear rate so that Equation 3.33 yields the same viscosity at any shear rate, then the liquid is called "Newtonian[11]" (see Figure 3.17).

Some liquids obtain a (gel like) structure when allowed to rest for some time. This structure must be overcome before the sample can flow again. The shear stress

[11] From Sir Isaac Newton

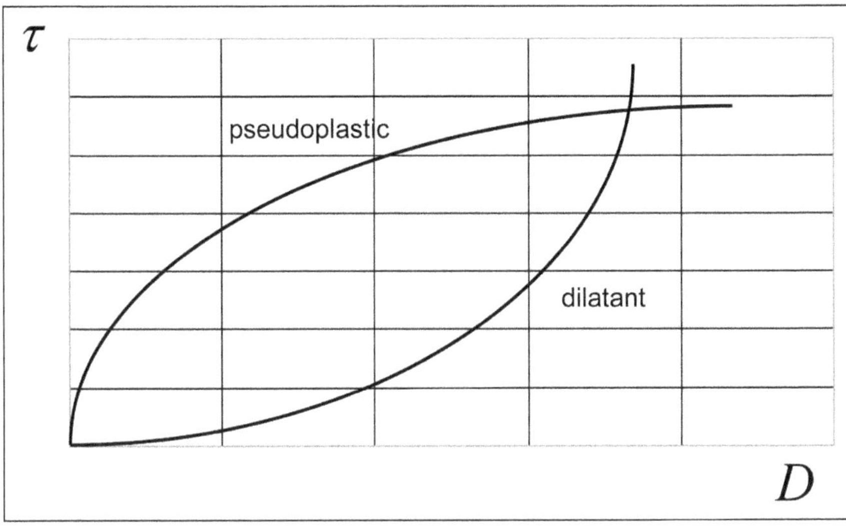

Figure 3.18: Pseudoplastic and dilatant flow behaviour

which is necessary for that to happen is called the "yield point". If a sample has such a yield point, but otherwise flows in a Newtonian fashion, it is called a Bingham liquid. Examples for this type of materials are tooth paste, ketchup and mayonnaise. The typical rheogram of a Bingham fluid is also shown in Figure 3.17.

The shear stress is not necessarily, or better, is seldom linearly dependent upon the shear rate. There are shear thinning and shear thickening systems, depending upon if the shear stress increases less than proportionally or more than proportionally with increasing shear rates. Accordingly, the viscosity either falls or rises with increasing shear rate. Shear thinning systems are called "pseudoplastic" or are described as systems with "structural viscosity" since their viscosity is dominated by an inner structure that is decomposed during shearing. The cause for pseudoplasicity can be the intermingling of polymer molecules or the interaction between suspended colloidal particles such as pigments.

Contrarily, in shear thickening systems, a structure is established during the flow. In that case the liquids are called "dilatant". Dilatency is for example found in cornstarch solutions. Highly filled suspensions of solid particles like bread dough, slurries of sand or highly filled pigment pastes also flow in a shear thickening manner. The reason for this is that the particles separate from one another during the flow and the voids that are generated have a higher liquid demand. The suspensions become "drier". A vivid example can be observed when walking on a sandy beach at the coastal line where the waves drench the sand. Fresh foot prints appear to be drier than the unperturbed sand next to them. Figure 3.18 shows the rheo-

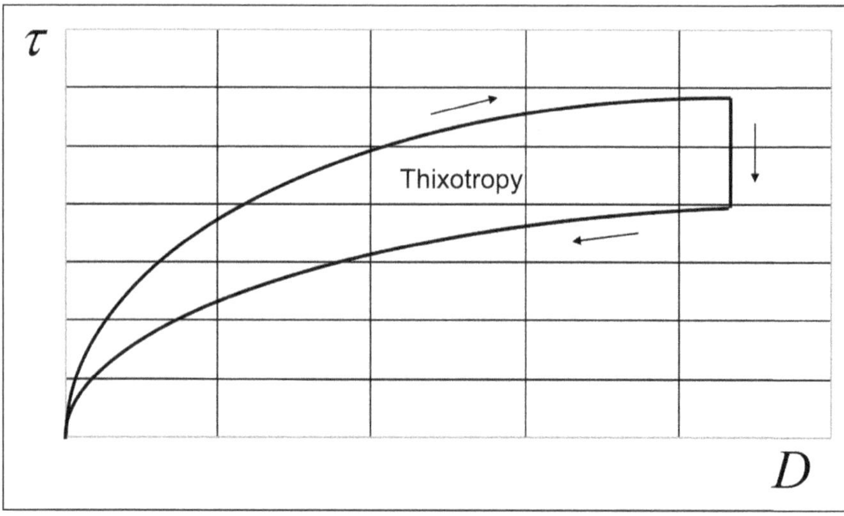

Figure 3.19: Thixotropic flow behaviour (arrows depict the chronological cycle of the experiment)

gram for pseudoplasitc and dilatant flow. In the case of Newtonain or Bingham fluids as well as in simple pseudoplastic or dilatant fluids (Figures 3.17 and 3.18), the shear stress is an explicit function of the shear rate. For each shear rate there is only one defined shear stress value, regardless if the liquid is sheared up or down.

In systems that show pseudoplastic or dilatant flow behaviour, the viscosity can also depend upon the time that the sample is sheared. In the case of time dependent shear thinning, one speaks of "thixotropy". Figure 3.19 shows a typical rheogram of a thixotropic liquid schematically. Such rheograms are obtained when a sample is sheared while continuously increasing the shear rate. At the highest shear rate chosen for the measurement, the shear rate is kept constant for some period of time. Because of a further decrease in inner structure, the shear stress (and the viscosity with it) drops. Then the sample is sheared down, i.e. the shear rate is gradually reduced to zero. The area that is surrounded by the flow curve is a measure for the thixotropic nature of the sample. Of course, it depends upon the parameters that were chosen for the rheological experiment.

Time dependent shear thickening is called rheopecticity. Figure 3.20 shows the typical flow curve of a rheopectic fluid. The experimental assessment of such a flow behaviour again involves shearing up to a defined shear rate, a holding period and, finally, shearing down. All non-Newtonian fluids can have a yield point τ_0 too. A general flow function is:

Equation 3.24 $$\tau = \eta \cdot D^n + \tau_0$$

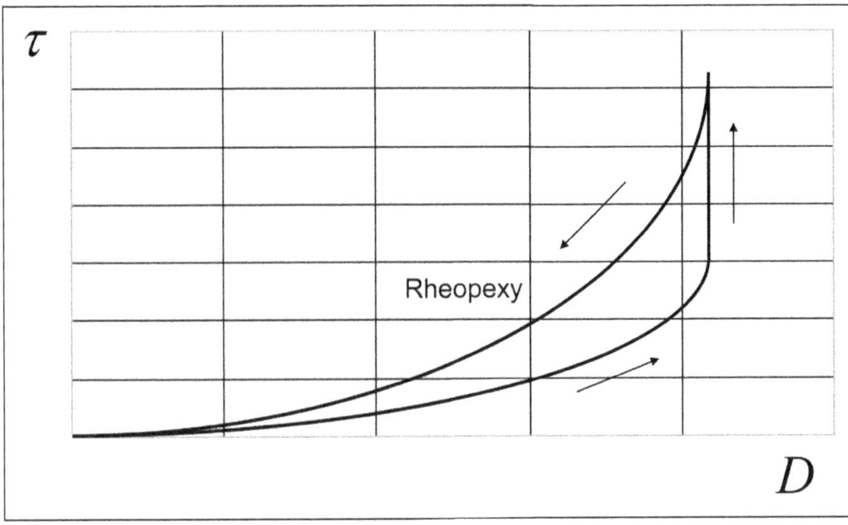

Figure 3.20: Rheopectic flow behaviour (arrows depict the chronological cycle of the experiment)

Table 3.6 shows all the different possibilities in a systematic way.

Viscosities are measured with rheometers. The liquid is placed into the gap between a carrier vessel and a rotor (Searle type of rheometer), or otherwise, a conus – plate or plate – plate geometry is chosen. "Controlled shear rate rheometers" are distinguished from "controlled shear stress rheometers" In the first type of instrument, the shear rate is applied and the resulting shear stress is measured. In controlled stress rheometer it is the other way around. In their case the rotor can charge a defined shear stress, whereas the resulting shear rate is measured. Further details can be taken from literature [41].

Table 3.6: Distinction of cases for the general flow function of Equation 3.24

$n = 1$	$\tau_0 = 0$	Newtonian fluid
$n = 1$	$\tau_0 > 0$	Bingham fluid
$n < 1$	$\tau_0 = 0$	Pseudoplastic fluid without yield point
$n < 1$	$\tau_0 > 0$	Pseudoplastic fluid with yield point
$n > 1$	$\tau_0 = 0$	Dilatant fluid without yield point
$n > 1$	$\tau_0 > 0$	Dilatant fluid with yield point

3.5.1 Mill base rheology

When formulating mill bases, two alternative routes are followed. Either the pigments are dispersed merely in a solvent (or solvents) in the presence of dispersing additives, or the mill base may also include resin. In the first case, the pigment loading tends to be higher, but the mill bases are mostly more abrasive and, due to the comparably higher prices of common dispersing additives, may be less economical. On the other hand, resins used for dispersion perform well in terms of wetting, dispersing and pigment stabilization against flocculation, but may have detrimental effects in the final, cured paint film.

Both the composition consisting of all the ingredients used and the temperature determine the flow behaviour of mill bases. The smaller the hydrodynamic volume of the resin molecules (that is the space filled by a swollen resin molecule) is, and the higher the temperature is, the lower the viscosity of its solution will be. Furthermore, the pigments play a role. In principle, the viscosity increases with increasing pigment loads. The increase is especially pronounced at higher pigmentation levels. The onset of viscosity increase is shifted to lower pigmentation levels in the case of smaller pigment particles and the viscosity increase is steeper. Figure 3.21 schematically shows the viscosity trend of pigment suspensions as a function of the pigment load ϕ.

In principle, mill bases with nearly Newtonian flow are desirable. Strong pseudoplasticity or thixotropy is just as harmful as dilatancy or rheopexy. In case of pronounced pseudoplasticity, the low viscosity hinders the transfer of shearing forces onto the agglomerates. When dilatant flow occurs, the machinery can even be caused to stop dead when the mill bases turn from liquid to solid within seconds. The only case where mild dilatancy is desirable is when dispersing with high speed impellers. More information on the required flow properties of mill bases for different dispersing machines is given in Chapter 4. In practice, pigment loads in mill bases are always so high that flow anomalies come by themselves. Typically, the pre-dispersed mill bases have a higher viscosity because resin solution is absorbed into the voids of the agglomerates so that it is depleted in the continu-

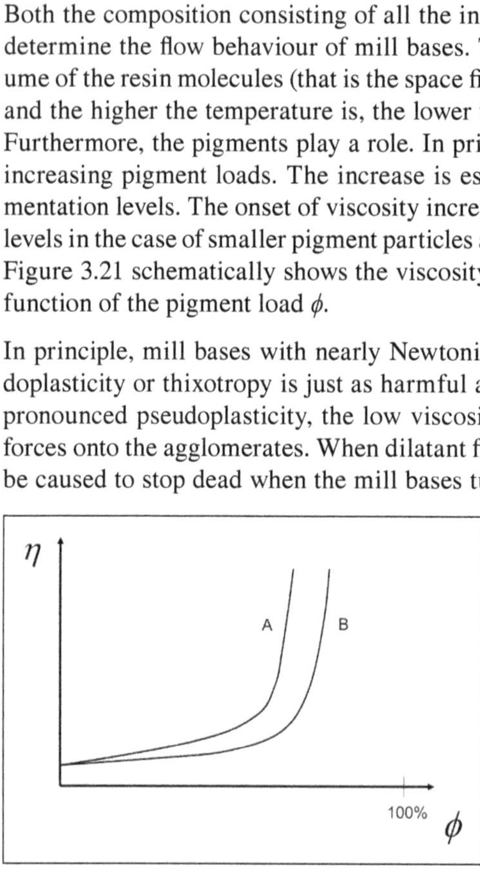

Figure 3.21: Schematic representation of the viscosity of mill bases in dependency of the volume fraction of pigment ϕ.
A: pigment with a smaller particle size
B: pigment with a larger particle size

ous liquid phase. When the mill base is dispersed, the agglomerates are caused to break and the absorbed resin solution is set free again so that the viscosity diminishes. Upon further dispersion, more primary particles are set free that have low separation distances to each other because of their low size and their high concentration. That's why in many cases the viscosity ascends again (see Figure 3.22).

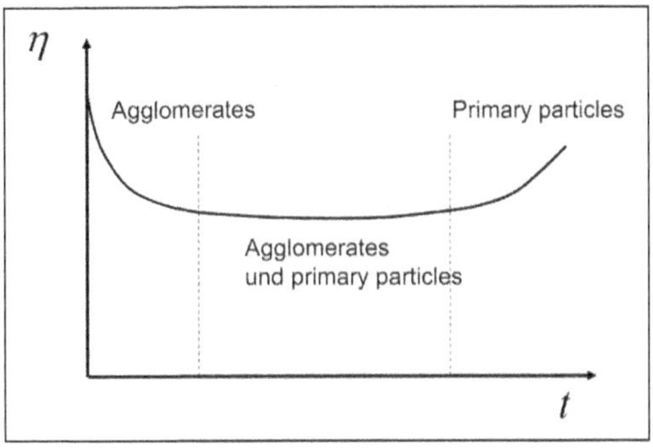

Figure 3.22: Schematic representation of the development of viscosity in the course of a dispersion. Dotted lines separate regions in which agglomerates predominate flow behaviour, in which agglomerates and primary particles are in existence next to each other and in which the influence of primary particles overrules.

Without surface active substances like wetting agents, dispersion additives and/or resins, high loaded mill bases cannot be prepared and, on many occasions, the viscosity of mill bases or finished paints can be lowered by the addition of dispersing additives. It was already mentioned that the viscosity of mill bases at equal pigment loads is higher, the smaller the pigment particles are. That is why inorganic pigments and fillers can normally be used in higher concentrations than organic pigments or carbon blacks. Mill bases of pigments are generally at least pseudoplastic, if not thixotropic. That is why the viscosity is lower at high shear rates than at low shear rates.

This is shown in Figure 3.23. Also shown is the influence of particle size on the flow behaviour. The lower the mean particle size is, the higher the viscosity at low shear rates will tend to be. Dispersions themselves are typically carried out at shear rates somewhere between 1000 and 1000,000 1/s [42]. At these shear rates, loose structures caused by particles touching each other will break down so that the flow behaviour no longer depends upon the particle concentration to such an extent. Although, if the particle concentration is too high for the amount of wetting agents used, the dispersed systems will tend

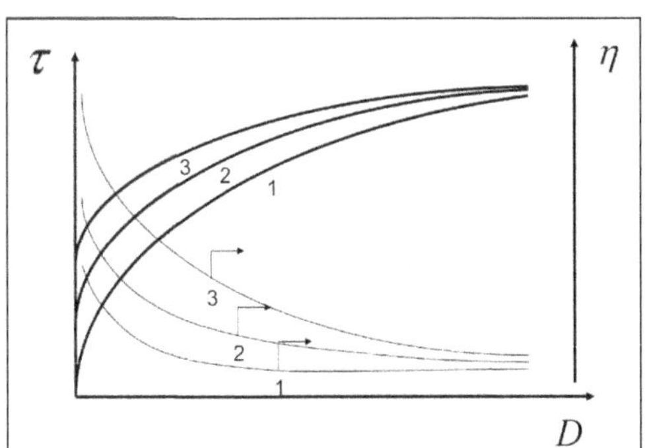

Figure 3.23: Schematic representation of the flow behaviour of pseudoplastic mill bases in dependency of the mean pigment particle size: $d_1>d_2>d_3$

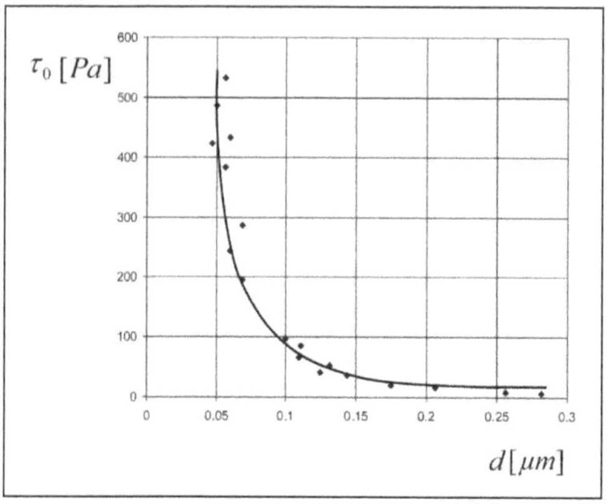

Figure 3.24: Yield points of Cl 15/16 – copper phthalocyanine pigment at a pigment volume concentration of 13 % (redrawn from [42])

to form yield points so that the pigment load is limited. Figure 3.24 illustrates schematically that yield points are higher and more probable, the smaller the pigments are. Figure 3.24 shows the dependency of the yield point on the mean particle size in the case of a chlorinated copper phthalocyanine pigment at a pigment volume concentration of 13 % in a resin solution (redrawn from [42]).

Instead of measuring the dynamic viscosities, the flow properties of paints and mill bases are often determined with flow cups. A flow cup is a beaker with a conical bottom which ends in a nozzle of defined size. The cup is first filled with the medium to be tested and then the nozzle is let free. The time the fluid needs leave the cup completely is measured with a stop watch. In the EN ISO 2431 from 1996, beakers with nozzle diameters of 3 mm, 4 mm, 5 mm and 6 mm are specified. However, there are numerous other standards such as ASTM D1200 (Ford cup), BS 3900; Part A6, 1971, AFNOR Cups, NF – T – 30014 or ASTM D 4212 (Zahn cups) with other sizes and nozzle diameters. Determining the flow time only makes sense if the fluids are not too far from Newtonian in their flow behaviour. In order to obtain meaningful results, the size of the cup should be selected so that the flow time is somewhere between 30 and 100 seconds and it must be ensured that the liquid flow breaks in a defined manner so that the end of the measurement can be determined reproducibly.

For Newtonian fluids, the flow time is connected to the kinematic viscosities. The kinematic viscosity is the quotient between the dynamic viscosity and the density of the liquid. The algorithms for transforming flow times from different flow cups into values for the kinematic viscosity are listed in EN ISO 2431, for example. For that, according to that standard, kinematic viscosities should not exceed a value of 700 mm^2/s.

3.5.2 Mill base optimization for bead mills; determination of binder demand

One method for characterizing the wetting properties of liquids is used to optimize mill bases. It is the measurement of the binder demand by determining the "flow point" according to Daniel [43]. In a wide sense, this method is analogous to the oil

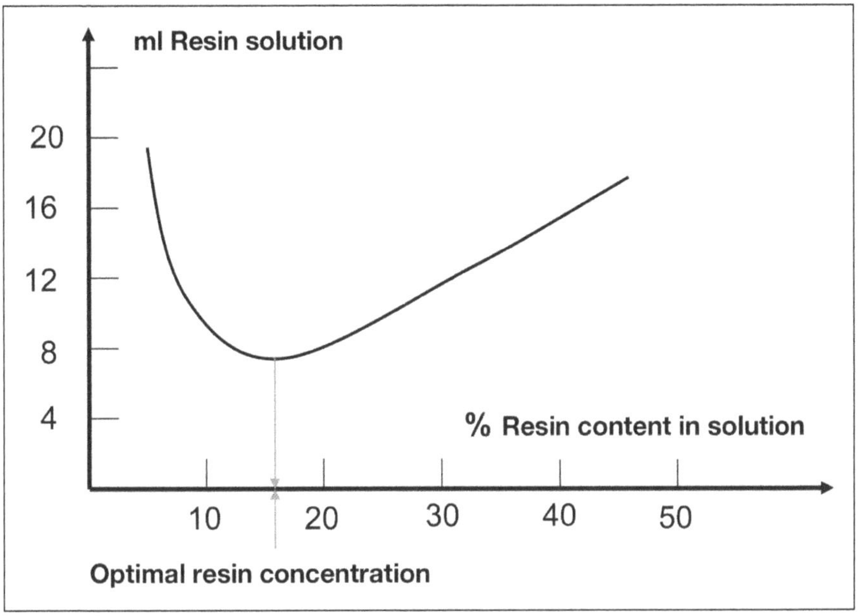

Figure 3.25: Mill base optimization through binder demand

absorption value (ISO 787, Part 5, 1980). The oil absorption value is a relict from times in which alkyds and polyester resins were almost solely used in coatings. These resins were modified ("boiled down") with fatty acids esters ("oils") so that the wetting properties of pigments towards, e.g., linseed oil was a meaningful value. Just about every industrialized country on the globe had its own standard for measuring the oil absorption. To determine the oil absorption number according to ISO 787, Part 5, the pigment is placed upon a glass plate and linseed oil with a specified acid number is added dropwise to the pigment from a burette. The pigment and oil are mixed with a spatula until a homogenous, mastic type paste is obtained. The oil absorption number is defined as the amount of linseed oil necessary to wet 100 grams of pigment in such a way. In contrast to that, the amount of oil necessary to maintain a free flowing paste which can be applied for dispersing in a piece of machinery is distinguished. The amount of oil necessary for that is called the "flow point". In most cases the flow point is reached at a rather precise oil amount. Yet, both the oil absorption number and the flow point have a relatively strong dependency upon the force used while mixing since, when kneading the mixture, pigment surface area becomes available for wetting as dispersion takes place. Both the oil absorption number and the flow point give information only related to the specified solid-liquid system under investigation. Instead of linseed oil, any resin solution can be used. Then, analogous to the flow point, the "binder demand" value, that is the volume of resin solution needed to reach the flow point, is gained. When binder demands are determined for differently concentrated resins solutions, different flow points will be found. The graphical plot of the flow points as a function of the concentration of the resin solution is often a skewed, parabolic type function with a more or less pronounced minimum at some resin concentration [44]. This minimum depicts the composition of the resin solution with the best wetting properties for that pigment, whereas the corresponding value on the ordinate shows the amount of resin solution needed (Figure 3.25).

The flow point method is best suited to optimize mill base compositions for bead mills. It is advisable to start off with 10% or 20% higher resin solution amounts than found from the flow curves, though (Figure 3.25). Otherwise, due to the efficient dispersing action of these machinery, more surface may be unclosed than in the flow point determination and the mill base may become dilatant or even solidify, causing the machine to stop abruptly.

3.6 Literature

[1] J. Winkler, Farbe und Lack 94 (1988) p. 263–269

[2] G. Wedler, Lehrbuch der Physikalischen Chemie, Wiley-VCH, 5. Auflage, Weinheim, 2004, p. 415 ff

[3] Handbook of Chemistry and Physics, CRC Press, 88th Edition, 2007–2008, p. 6–147

[4] T. Young, Philosophical Transactions of the Royal Society of London, The Royal Society, London, 95 (1805) 65–87

[5] M. E. Schrader, J. Adhesion Sci. Technol. 6, No. 9 (1992) p. 969–981

[6] J. B. Donnet, C. M. Lansinger, Kautschuk + Gummi Kunststoffe, 45, Nr. 6 (1992) p. 459–468

[7] W. A. Zisman, Journal of Paint Technology, 44, No. 564 (1972) p. 42–57

[8] H.-J. Jakobasch, "Oberflächenchemie faserbildender Polymerer", Akademie Verlag, Berlin 1984, p. 85–87

[9] A. Horsthemke, J. J. Schröder, Chem. Eng. Process. 19 (1985) p. 277–285

[10] W. A. Zisman, Advances in Chemistry, Series 43, (1964) p. 1–51

[11] D. M. Gans, J. Paint Technology, 41, No. 536 (1969) p. 515–522

[12] L. A. Girifalco, R. J. Good, J. Phys. Chem. 61 (1957) 904; ibid. 64 (1960) 561

[13] F. M. Fowkes, Ind. Eng. Chem. 56 (1964) p. 40–52

[14] F. van Voorst Vader, Chem. Ing. Techn. 49, Nr. 6 (1977) p. 488–493

[15] D. K. Owens, R. C. Wendt, Journal of Applied Polymer Science 13 (1969) p. 1741–1747

[16] D. K. Owens, R. C. Wendt, W. Rabel, farbe + lack 77 (10) (1971) p. 997–1005

[17] S. Wu, J. Polymer Sci., Part C, No. 34 (1971) p. 19

[18] S. Wu, J. Phys. Chem., 74 (1979) p. 632

[19] A. W. Neumann, P.-J. Sell, Z. Phys. Chemie, Bd. 227, Heft 3/4 (1964) p. 187–193

[20] H. Schubert, Chemie Ing. Techn. 41 (1969) p. 1276

[21] P. M. M. Vervoorn, Colloids and Surfaces, 25 (1987) p. 145

[22] H. Schubert, Chemie Ing. Techn. 40, Nr. 15 (1968) p. 745–747
[23] F. E. Bartell, H. J. Osterhoff, Ind. Chem. Phys. 19 (11) (1927) 1277
[24] E. W. Washburn, Phys. Review 17 (1921) 273
[25] B. Etmanski, A. K. Bledzki, U. Fuhrmann, Kunststoffe 84 (1994) p. 46–50
[26] S. Wu, K. J. Brzozowski, Journal of Colloid and Interface Science, 37, No. 4 (1971) p. 686–690
[27] N. W. F. Kossen, P. M Heertjes, Chemical Engineering Science 20 (1965) p. 593–599
[28] J. T. Fell, E. Efentakis, Int. J. Pharm. 4 (1979) p. 153
[29] G. Buckton, Powder Technol. 46 (1986) p. 201–208
[30] U. Bröckel, F. Löffler, Part. Part. Syst. Charact. 8 (1991) p. 215–221
[31] C. F. Lerk, M. Lagas, Acta Pharmaceutica Technologica, 23 (1971) p. 21–27
[32] D. W. Fuerstenau, M. C. Williams, Part. Charact., 4 (1987) p. 7–13
[33] J. B. Donnet, C. M. Lansinger, Kautschuk +Gummi Kunststoffe, 45, Nr. 6 (1992) p. 459–468
[34] S. Wolff, E.-H. Tan, J. B. Donnet, Kautschuk + Gummi Kunststoffe, 47, Nr. 7 (1994) p. 485–492
[35] G. M. Dorris, D. G. Gray, Coll. Int. Sci. 77 (1980) p. 353
[36] G. M. Dorris, D. G. Gray, Journal of Colloid and Interface Science 71 (1979) p. 93
[37] J. Winkler, “Gaschromatographie an Pigmentoberflächen”, Diplomarbeit, Universität Stuttgart, 1980
[38] M. v. Stackelberg, K. H. Frangen, Forschungsberichte Wirtschafts- und Verkehrsministerium Nordrhein-Westfalen, 166 (1955) p. 47–49
[39] F. Kindervater, Deutsche Farben Zeitschrift 14 (1960) p. 49–53
[40] F. Kindervater, Farbe u. Lack 69 (1963) p. 21–26
[41] T. Mezger, “The rheology Handbook”, Vincentz Network, 2011
[42] J. Schröder, Rheologie 92 (1992) p. 40–48
[43] F. K. Daniel, Official Digest 28, No 381 (1956) p. 837–857
[44] H. Rohrer, Farbe und Lack 69 (1963) p. 591

4 Dispersing equipment

The mechanical breaking of pigment agglomerates is performed within dispersion machines. In dependency of the mill base viscosity and the aspired quality of the dispersion, different machines may be considered. Apart from that, other aspects such as wear of the machinery or de-aerating properties may influence the choice of dispersion equipment. Sometimes different machines are utilized consecutively, for example a first dispersion with a high speed impeller followed by a fine dispersion with a bead mill. The most important features of

- High speed impellers
- Three roll mills
- Kneaders and extruders and
- High speed attritors (bead mills)

are discussed in this chapter and recommendations for optimum operation conditions are given.

4.1 High speed impellers

A high speed impeller is a discontinuously operated piece of dispersion machinery. The mill base is placed into a (in the case of laboratory machinery temperature controllable) vessel equipped with a rotating shaft that ends in a toothed disk (see Figure 4.1)

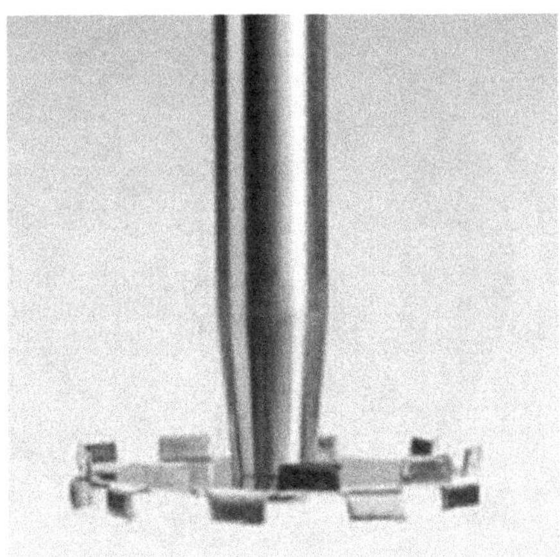

Figure 4.1: Impeller blade on the shaft of a high speed impeller
Source: VMA Getzmann GmbH, Reichshof-Heienbach, Germany

The teeth on the impeller blade are alternately bent upwards and downwards and are slanted relative to their driving direction in such a way that the mill base is transported circularly around the shaft as well as tangentially outwards. The inclination of the teeth from the driving direction should lie between 11° and 30° [1].

Jochen Winkler: Dispersing Pigments and Fillers

ISBN 978-3-86630-852-7

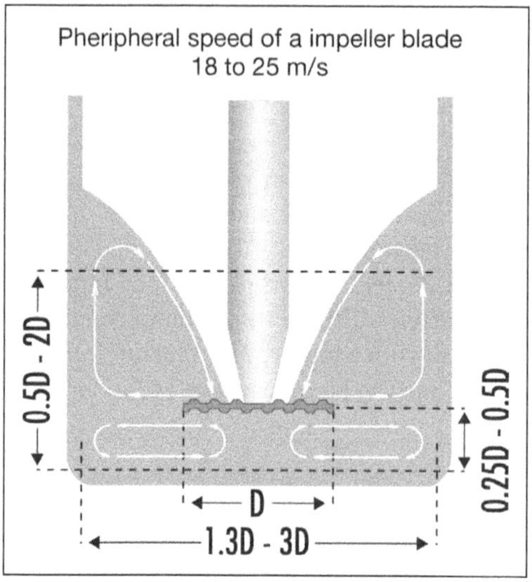

Figure 4.2: Dimensions of a high speed impeller and flows within the dispersion vessel
Source: VMA Getzmann GmbH, Reichshof-Heienbach, Germany

Figure 4.3: Doughnut effect

When in use, a liquid flow is generated in the dispersion vessel as shown schematically in Figure 4.2. While reaching the vessel wall, the mill base separates into an upper and a lower annular-shaped stream. Figure 4.2 gives an indication of some relative size dimensions which have proven to be beneficial in practice. The sizes are all related to the diameter D of the toothed impeller blade.

The filling level of the mill base should be somewhere between one and a half and twice the impeller blade diameter. The rotational speed should be adjusted so that the impeller blade becomes visible from the top and that there are no places near the wall of the vessel where the mill base is at rest. The annular flow that is generated is named “doughnut effect” after the well-known American pastry (Figure 4.3).

The dispersing effect arises to a lesser degree in the vicinity of the teeth at the rim of the impeller blade, for example by agglomerates being hit directly by them. On the other hand, a shear gradient is formed within the mill base, whereby the liquid on one side of an agglomerate flows at a higher speed than on the opposite side. As will be discussed in Chapter 5, the dispersing action of a shear force is more pronounced, the higher both the shear rate and the higher the shear stress within the suspension is. That is why high speed impeller dispersions are run best at possibly high viscosities. This is in agreement with the views of many authors [2-5] that recommend high viscosities up to mildly dilatant mill base flow properties.

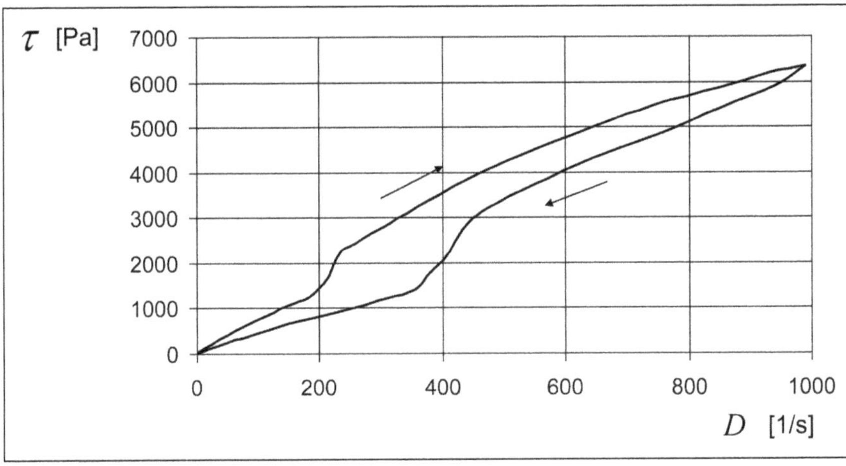

Figure 4.4: Rheogram of a mill base with a dilatant flow (synthetic barium sulfate with a mean particle size of 0.7 μm)

Rheograms (viscosity curves) of highly pigmented mill bases often exhibit a dilatant step in a certain shear rate interval. Dilatancy characterizes a flow behaviour in which the shear stress increases more than proportional to the shear rate so that the viscosity also increases more than proportionally. As an example, Figure 4.4 shows the rheogram of a mill base of a synthetic barium sulfate with a mean particle diameter of 0.7 μm. In the shear stress versus shear rate curve, there is a distinct dilatant step between 200 s^{-1} and 300 s^{-1} while shearing upwards. In the downward shearing part of the rheogram, dilatancy is observed between 350 s^{-1} and 450 s^{-1}. The advantageous influence of dilatancy is probably created by the effect, that the agglomerates are hindered in their rotation, whereby the shearing action is improved. After dispersion, the mill base no longer showed dilatancy, at least not in the shear rate range in which the measurement took place.

There is consensus that the viscosity of the resin solution in the mill base should not be too high. *Ensminger*[5] recommends that the resin solution in the mill base should have a viscosity between 50 and 100 mPas (milli-Pascal seconds). Pigment should be added to the resin so that a viscosity of approximately 3000 mPas to 4000 mPas is reached. Care should be taken that the resin solution is not too lean. To avoid problems in the let-down, the resin solids content should be at least about 15 weight per cent. At too low viscosities in the mill bases, air is easily entrapped and the dispersing effect is furthermore inefficient because of low shear. Figure 4.3 shows the flow pattern in a correctly formulated mill base.

High speed impellers are normally operated under the aspect of a certain circumference velocity at the tip of the blade. This should lie somewhere between

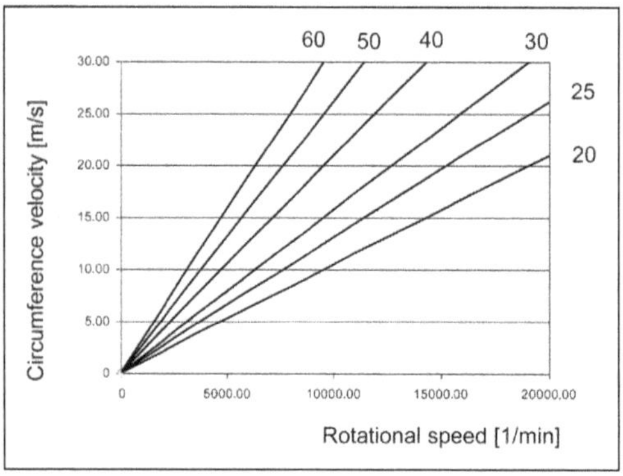

Figure 4.5: Circumference velocities of different impeller blades in dependence of their diameters and the rotational speed

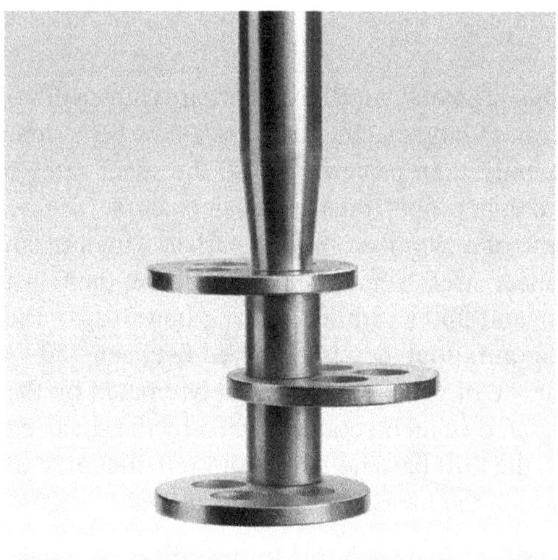

Figure 4.6: Bead mill inlet
Source: VMA Getzmann GmbH, Reichshof-Heienbach, Gemany

18 and 25 meters per second. Especially in the case of laboratory equipment, i.e. with small impeller blades, this implies rotational speeds up to approximately 20,000 revolutions per minute. Figure 4.5 shows the rotational speeds necessary to obtain the circumference velocities using different impeller blade sizes.

Apart from being used for pre-dispersing paints, high speed impellers are often used in the production of pasty media such as SMCs (SMC, sheet moulding compounds) or glues. Especially in highly viscous media, de-aeration can present a problem because the bothersome, small air bubbles do not have enough buoyancy to rise to the surface and leave the system. That's why machinery was developed in which the dispersion vessel can be exposed to vacuum. The vacuum sucks the entrapped air out of the agglomerates and lets the bubbles rise to the surface where they are removed. Furthermore, no air can be stirred into the mill bases, simply because it's not present. When operating such a piece of equipment, the vacuum must be controlled so that solvents do not evaporate.

In the laboratory it is possible to use high speed impellers as open discontinuously operated bead mills. For this, discs of ceramic, metal or a tenacious plastic are mounted instead of the impeller blade (see Figure 4.6). Milling beads are filled into the dispersion vessel and, upon operation; the mill base is added so that the bed of beads is just covered.

4.2 Roller mills (three roll mill)

Three roll mills (see Figure 4.7) are dispersion equipment that unfold their dispersing action merely by shearing of the pigment agglomerates. They consist of an arrangement of three cylindrical rolls, a feed roll, a centre roll and an apron roll, that rotate in this order with progressively higher speed. The mill base is applied between the feed roll and the centre roll. It enters the gap between them and is sheared. Agglomerates which are larger in size than the gap are compressed. Due to the higher rotational speed of the centre roll, the mill base adheres to the centre roll more than to the feed roll so that it is passed on to the centre roll. For the same reason, the mill base then goes over to the apron roll. The dispersed mill base is collected by a knife blade that scrapes the dispersed formulation from the apron roll.

In most cases, the pressure with which the rolls are pressed against each other can be adjusted. Without a mill base feed, there are no nips between the rolls. The mill base itself squeezes between the rolls and lets a gap form which allows it to pass through. The nip size depends upon the applied pressure in a straight forward manner. However, machinery is also commercially available in which the gap size can be adjusted directly and very precisely. A shear gradient is generated that is determined by the interaction of gap width and the relative rotational speed of the rolls.

Three roll mills are mainly used to produce printing inks. Their advantage is that, next to a good and uniform particle distribution, they de-aerate the inks so that small air bubbles do not create pin holes or craters after application. Since the mill bases are handled openly, the emission of solvents has to be considered from an industrial hygiene point of view. There are stringent requirements on the dimensional accuracy of the rolls since, otherwise, the gap would not be uniform in width.

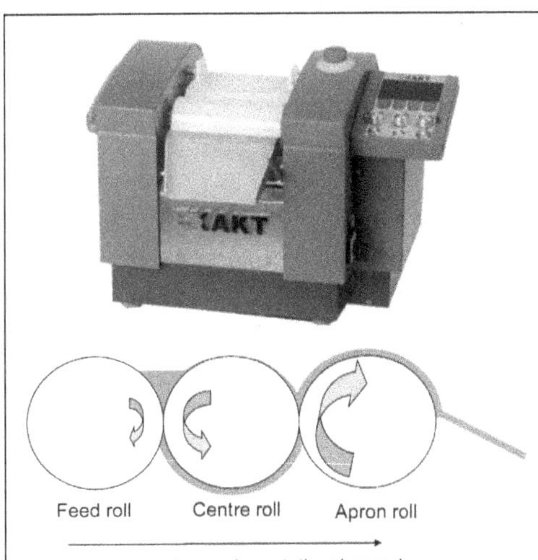

Figure 4.7: Three roll mill

As far as dispersion results are concerned, a high mill base viscosity is beneficial. Yet, it must be guaranteed that the mill base forms a continuous film on the rolls, that it can be collected from the apron rolls and that it then flows into a collector vessel by itself. More and detailed information on the influence of roll dimensions, their rotational speed, mill base viscosity and the gap width on the mill base discharged and the energy uptake of this type of machinery as well as strategies for determining the rotational speeds of the three separate rolls for optimum dispersion results is found in literature [6].

4.3 Kneaders and extruders

Kneaders and extruders (Figure 4.8) are both dispersion equipment in which the mill bases are sheared on immobile surfaces. They are used for the production of dispersions in very viscous media. In kneaders, internal mixing tools are used to shear the feed. A typical product manufactured in kneaders is, for example, filler material like spackling paste.

In extruders, one or more agitating screws turn within a tube, in principle like in a meat mincer. Extruders may be heated along the path of the screw so that pigments and fillers can be worked into plastic melts. Different zones are distinguished in an extruder. In the feed section the powdery educts: thermoplastic resin, pigments and fillers, are brought into the machine, mixed and heated. This is followed by the plasticizing zone in which the polymer is molten. When producing powder coatings, temperatures of 80 °C to 100 °C should not be exceeded since, otherwise, the curing reaction could commence. Due to a decline in the duct volume, the mass is compressed and possible air filled voids are destroyed.

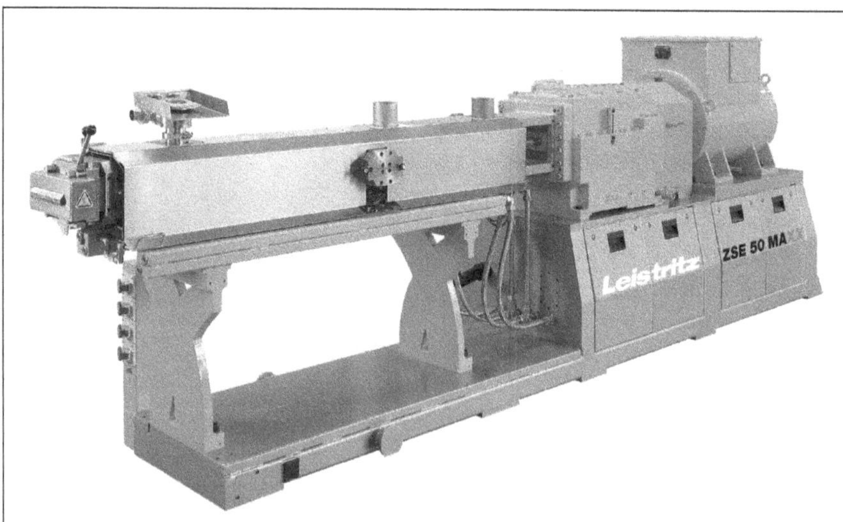

Figure 4.8: Extruder

The air leaves the machine in the direction of the feed. The screws transport the melt onwards to the homogenizing zone, where further dispersing and mixing takes place. The screws are constructed in a manner that the melt is consecutively compressed and expanded, thus generating a shearing action. By applying vacuum, volatile ingredients such as rest monomers, moisture or products from cracking reactions can be extracted.

The molten plastics that leave the extruders are formed into strands or flat sheets that are cooled until they turn solid. In a further step, the strands or sheets are granulated by cutting or breaking. In plastic industry, “master batches” are produced this way. Master batches are granules with a high amount of dispersed pigment in them that are later used in injection moulding by mixing them with granules of the pure polymer. Needle shaped pigments can break in extruder dispersions.

Most machines allow for measuring the torque of the screws so that the mechanical power brought into the systems can be determined[1]. It must be taken into consideration, though, that by increasing the rotational speed in order to increase the mechanical power, the dispersion time is lowered since the screws discharge the melts faster. If the degree of deagglomeration is not improved or even becomes worse, then it makes more sense to increase the power input by extruding at lower temperatures where the melt will be more viscous and lead to a higher torque of the shaft.

In extruders, the agglomerates may be dispersed either by erosion or they may break spontaneously and completely under the shear stress. Erosion is related to the separation of primary particles from the agglomerate surface, just like a snowball that rolls on a street. Both dispersion mechanisms were investigated and expressed mathematically [7]. They are different in a sense that erosion may occur at lower powers than needed for the spontaneous breakage of agglomerates since at a given time only few (ideally only one) physical bonds have to be split. Below a critical power that is necessary for complete breakage in one step, only erosion can take place at best. The decrease in agglomerate size per time unit is far less in erosion than in spontaneous deagglomeration [7]. In both cases, agglomerate size reduction is directly proportional to the product between the shear stress τ_L and the shear rate D in the mill base as well as to the mean diameter d_A of the agglomerates and inversely proportional to the agglomerate strength τ_A. In the case of erosion, the decrease in agglomerate size also depends upon their porosity ε_A. The porosity is the fraction of void volume of an agglomerate, so it is the quotient between the void volume and the volume of the complete agglomerate. For $\vartheta d/\vartheta t$, the reduction in agglomerate size per time unit, one can write for spontaneous breakage

[1] The mechanical power is the product of the rotational speed and the torque of the screw, multiplied by the factor 2π. It has the physical dimension Watt.

Equation 4.1
$$\frac{\vartheta d}{\vartheta t} = \frac{5}{128 \cdot \pi} \cdot \frac{\tau_L \cdot D}{\sigma_A} \cdot d_A$$

whereas in the case of erosion the agglomerate size depletion may be expressed by:

Equation 4.2
$$\frac{\vartheta d}{\vartheta t} = \frac{5}{256 \cdot \pi} \cdot \frac{\tau_L \cdot D}{\sigma_A \cdot (1 - \varepsilon_A)} \cdot d_A$$

So, agglomerates are dispersed more quickly the larger they are, the higher the shear rate and the shear stress in the mill base are and the lower the agglomerate strength is. For erosion to take place, it is advantageous if the agglomerates are very porous.

Since on one hand the void volumes normally are in the order of 0.5, yet on the other hand in Equation 4.2 there is a factor of 5/256 instead of 5/128, one might expect the agglomerate size reduction to be similar for both mechanisms. However, a number of different cases have to be considered.

1. In the case that the power is lower than necessary for erosion to take place, nothing is dispersed. Contrarily, agglomeration happens because of the adhesion (van der Waals attraction) between the agglomerates that meet.
2. When the power introduced into the system is larger than the critical power necessary for erosion to occur, but lower than the critical power for complete breakage, then only erosion takes place according to Equation 4.2.
3. If however, in the third case, the power is larger than would be essential for complete breakage, then both erosion and spontaneous deagglomeration happen. The resulting depletion of the agglomerate size is given by the sum of Equations 4.1 and 4.2. Size reduction is then about twice as fast as when only erosion occurs.

4.4 Bead mills (high speed attritors)

High speed attritors, commonly known as bead mills, are continuously operated dispersion machinery in which a mill base is pumped through a chamber ("milling chamber") which is filled with "milling beads" ("grinding media") that are kept in movement by a stirred shaft. Dispersing occurs between the beads that glide past each other. Figure 4.9 shows a cross section of a bead mill.

Bead mills indirectly derive from ball mills (more or less historical) in which cylindrical jar shaped vessels filled with mill base and grinding balls are turned. In these machines the grinding media rise along the rotating inner wall of the milling vessel and either cascade down upon the mill base or just slip off the walls. Today, these machines are no longer used for producing paints, except maybe on a laboratory scale. However, they are still run for wet and for dry

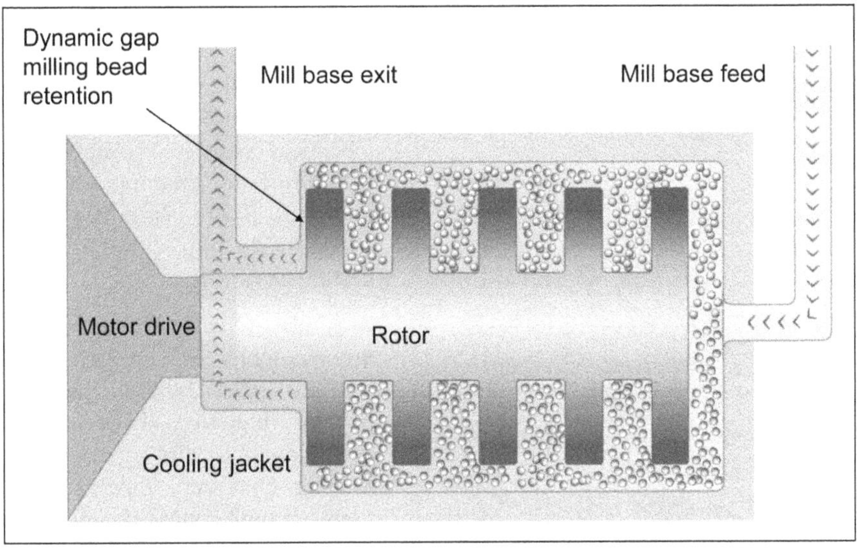

Figure 4.9: Cross section of a bead mill

grinding of ores or calciner discharges in titanium dioxide pigment production. The disadvantage of ball mills is their restricted rotational speed as well as the possibility for grinding media not to even separate from the wall of the vessel but, due to centrifugal force, sideslip.

Likewise outdated, at least for what making paints is concerned, are sand mills. These are open vessels filled with fine sand as grinding media. A stirred shaft keeps the bed of sand in a slow motion while mill base is pumped from the bottom of the vessel to the top. In former times, Ottawa sand, coming from a deposit near Ottawa in the State of Illinois, USA, was popularly used. It consisted of smoothly rounded, pure SiO_2 particles. These devices were able to produce very good dispersion results, although pre-wetting of the pigments some time before the actual dispersion ("ponding") aided their performance to some extent. However, due to the open construction, there are restrictions concerning the flow rate of the mill base and also its viscosity because of sand being carried out at the top. Yet, until this day, sand mills play a role in the production of inorganic pigments where they, for example, are used to grind the burner discharge from the chloride grade titanium dioxide production in aqueous slurries of low viscosity. Although today, zirconium silicate beads are used instead of sand because of its higher density.

Today's bead mills used for dispersing are closed systems in which the milling beads are retained in the milling chamber independent of the flow of the mill base by separation units. For those reasons almost any or at least very high

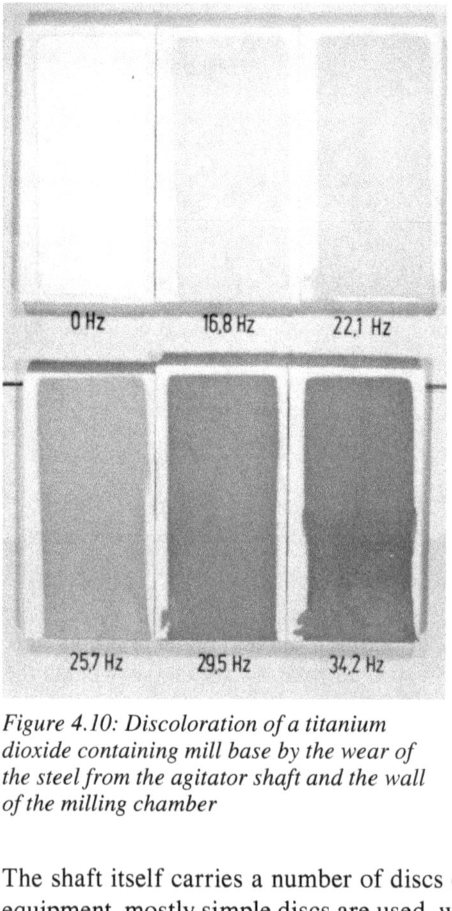

Figure 4.10: Discoloration of a titanium dioxide containing mill base by the wear of the steel from the agitator shaft and the wall of the milling chamber

rotational speeds are possible and the viscosity may vary within a wide range. Bead mills can therefore also be used for real grinding (chemical bonds are broken) of soft fillers like, for example, calcium carbonate. There is a great deal of literature dealing with these processes [8-10] [2].

A number of different techniques are available for separating the milling beads. Dynamic gaps are narrow interstices between the revolving shaft and a part of the milling chamber. Sieve separations are built somewhat less complicated since they are rigid. They can be located on a part of the milling chamber or on the shaft, so that the mill base can be discharged through the shaft itself. An advantage of this configuration is that bead crowding near the outlet is less probable because of the centrifugal force that drives the milling beads away from the shaft.

The shaft itself carries a number of discs of different geometries. In laboratory equipment, mostly simple discs are used, whereas in mills used for production or on a technical scale, discs are often constructed so that they convey the mill base and the milling beads away from the outlet to reduce bead crowding. Depending upon the milling task, different construction materials are used for the milling chamber and the shaft. The main issue of concern is to avoid the discoloration of sensitive milling media. Figure 4.10 shows an example for the influence of wear on the colour of a white pigmented mill base as a function of the rotational speed of the agitator shaft, leaving all other parameters constant.

Older bead mills used for production often have a vertical arrangement of the milling chamber, whereas in newer machinery, the milling chambers are preferably assembled horizontally. Although this has no influence on dispersion efficiency,

[2] It has already been pointed out that dispersing and grinding are two principally different processes, requiring different approaches. For more information see Chapter 5.

the initial torque when starting the machine tends to be lower because the discs do not reach completely into the bed of beads. Furthermore, the milling chambers are better accessible, for example for cleaning or maintenance purposes.

Due to high power inputs by the machines, the mill bases can heat up appreciably. Therefore the machines are always equipped with jacketed milling chambers through which cooling water is run. Depending upon the type of machine, the agitator shaft is also cooled to make sure that the mill base does not gather too much heat. Paints normally are able to withstand temperatures up to about 50 °C (approx. 120 °F) as long as this thermal load does not last too long. Temperatures higher than that may cause mill bases to gel. Disrupture of resin molecules (degradation of molecular weight) is theoretically possible, but does not present a problem in practice.

4.4.1 Milling beads

The type of milling beads used has a great influence on the dispersion efficiency. Milling beads are available from different materials, starting from simple glass beads over different ceramic beads up to beads made of steel. The basic quality criteria hereby are the density of the beads, their size distribution, their shape and, finally, their ability to withstand abrasion. Dispersing efficiency becomes better with increasing bead density because the kinetic energy of the beads is proportional to their mass (= weight). Table 4.1 was put together from brochures of milling bead producers. It lists the most important properties: density, bulk density (of a bed of beads in the size range between 1.4 and 1.6 mm diameter) and the Vicker's hardness of the materials. The beads are arranged in the order of increasing density. Glass beads have only low densities of little more than 2 g/cm³, whereas tungsten carbide features a density of 15 g/cm³.

Zirconium dioxide beads stabilized by different doping components have gained importance. Pure zirconium dioxide exists at room temperature as a monoclinic crystal with a density of 5.556 g/cm³. Upon heating, the crystal, in dependency of its size, is converted into tetragonal zirconium dioxide of the density 6.1 g/cm³ at temperatures between 950 °C and 1200 °C and, when heated further to 2370 °C, turns into cubic zirconium dioxide with a density of 5.83 g/cm³. Zirconium dioxide melts at 2600 °C. If the melt is allowed to solidify in the presence of MgO, CaO, Y_2O_3 or CeO_2, then cubic and tetragonal crystals of higher density are stabilized.

From experience, glass beads for example wear about 300 times faster than zirconium oxide milling beads. Wear of beads is disliked since the abraded material or, worse even, fragments of fractured beads with dimensions smaller than the separation units, are found in the milled product. On the other hand, milling beads loose their efficiency when they are worn to a lentil form. Worn beads of poor quality sometimes show that at least a part of the beads were hollow originally. Hollow spheres are less tough than solid beads and are handicapped with a lower density. Fresh milling beads are normally shiny, irrespective of their colour. In operation the beads become matt due to a roughening of the surface, but this has no effect on their dispersion efficiency.

Grinding beads are offered in mean particle sizes down to 0.05 mm (50 μm). Very fine beads are necessary to disperse nano-scaled agglomerates, since they form a dense bed in which the agglomerates are sheared in spite of their low inertia. The mean surface to surface distance of the beads in use is directly proportional to their diameters.

The bulk of milling beads should fill about 80 to 90 % of the free volume of the milling chamber. Lower loads lead to extremely poor performance, as can be seen in Figure 4.11, whereas higher loads promote bead wear as well as wear of the machinery. On top of that, very high filling levels inhibit the mobility of the beads so that dispersion efficiency is not improved further (compare Chapter 5.1 for the meaning of "colour strength").

Table 4.1: Physical properties of commercial milling bead types

Type	Density [g/cm³]	Colour	Buld density 1.4 - 1.6 mm	Vickers hardness [N/mm²]
Borosilikate glass	2.20	transparent	1.30	450
Soda lime glass	2.50	transparent	1.47	400
Silicium carbide	3.10	black	1.78	2300
Silicium nitride	3.20	gray	1.85	1400
Aluminum oxide	3.50	white	2.00	1600
Zirkonium silicate	4.55	white	2.30	700
Zirkononium oxide, silicium-stabilized	4.60	white	2.70	870
Zirkonium oxide, magnesium-stabilized	5.73	white	3.30	1000
Zirkonium oxide, yttrium-stabilized	6.05	white	3.70	1300
Zirkonium oxide, cerium-stabilized	6.20	braun	4.00	1250
Steel	7.80	metallic	4.80	200
Tungsten carbide, cobalt-stabilized	15.00	gray	-	1500

Bead mills can be classified as "ideal mixers". This means that the mill bases by no means pass the milling chamber in a plug flow. Due to the intensity of the stirring, there is a heavy axial mixing within the milling chamber. Any fresh mill base that enters the chamber is immediately mixed thoroughly with all the medium already present there and is then successively displaced by incoming mill base. The result of this is that the most frequent dwell time is significantly lower than the calculated mean dwell time. Figure 4.12 illustrates this effect. A dispersed white paint containing titanium dioxide as a pigment was run through a pin disc bead mill (PM-STS1 Draiswerke, Mannheim, Germany).

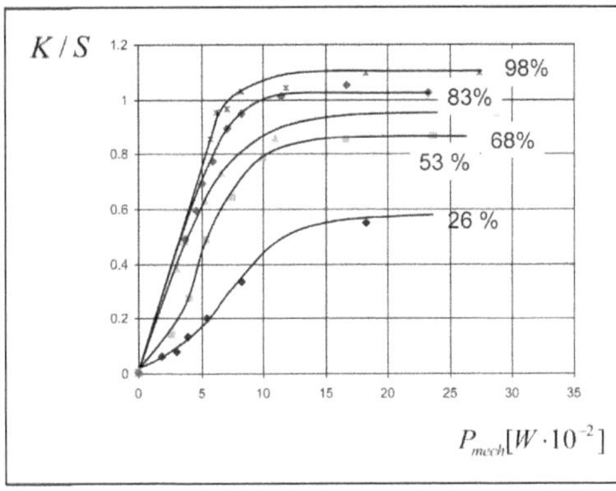

Figure 4.11: Influence of the milling bead load on the development of colour strength (as a measure for the degree of de-agglomeration) as a function of the mechanical power introduced into the milling chamber

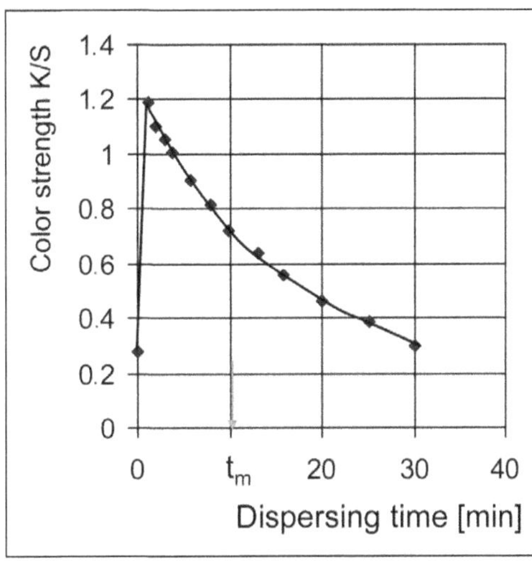

Figure 4.12: Experiment for determining the dwell time distribution in a 1.4 litre pin disc mill. The mean dwell time was calculated to be 10.1 minutes. For details, see text.

At the time zero, a small amount of a dispersed coloured pigment paste was added to the feed of the milling chamber by a syringe. Paint samples were drawn at progressing times and their colour strengths were calculated from the Kubelka Munk formula (see Chapter 5.1). Although the mean dwell time in the experiment was 10.1 minutes, the highest colour strength was found after less than one minute already. This proves that a large number of agglomerates only remain in the milling chamber for only a few seconds time. One may further deduct from this experiment that there is no temperature gradient within the milling chamber. The temperature is the same anywhere within the milling chamber and identical to the temperature

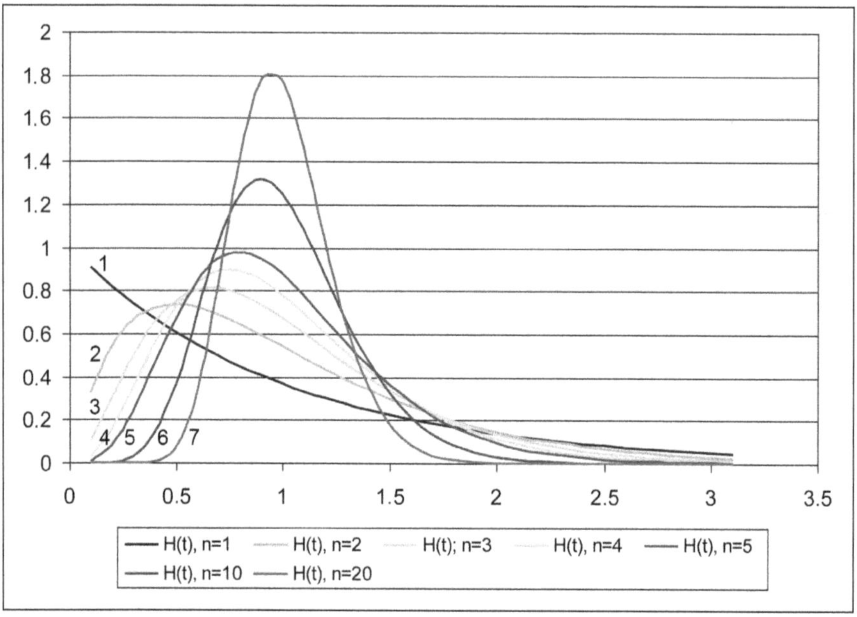

Figure 4.13: Dwell time distribution in n serially driven ideal mixers

of the mill base leaving the mill. Only in large mills or in mills where discs separate the milling chamber into different sections, thus preventing axial mixing, is a temperature gradient likely to occur.

It can be taken from common textbooks of chemical engineering that a near Gaussian distribution of the dwell time is only to be expected after about ten consecutive runs through ideal mixers [11]. The logical reason for this is that the probabilities for agglomerates to stay in the milling chamber for the same time at each pass decreases with the number of passes. The frequency $H(t)$ of the dwell time t is a function of the theoretical mean dwell time t_m and the number of passes n and can be calculated from Equation 4.3

Equation 4.3
$$H(t) = \frac{n^n}{(n-1)!} \cdot \left(\frac{t}{t_m}\right)^{n-1} \cdot exp - \left(\frac{n \cdot t}{t_m}\right)$$

In turn, the theoretical mean dwell time is found as the quotient between the free volume, that is the volume in the milling chamber free of milling beads V_{free}, and the volume flow of mill base $\dot{V}_{MB}$

Equation 4.4
$$t_m = \frac{V_{free}}{\dot{V}_{MB}}$$

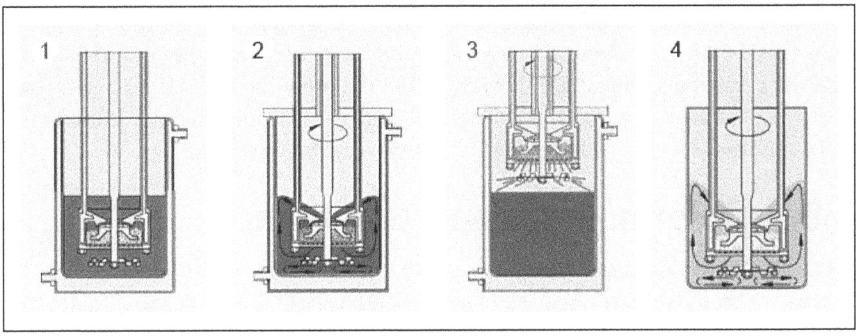

Figure 4.14: Operation of a basket mill; for explanations, see text
source: VMA Getzmann GmbH, Reichshof-Heienbach, Germany

The calculated mean dwell time distributions of ideal mixers after 1, 2, 3, 4, 5, 10 and 20 consecutive passes is shown in Figure 4.13. For better apprehension the abscissa in Figure 4.13 is plotted as the quotient t/t_m, so that the theoretical mean dwell time is at the value 1.

The balancing of heat flows in a high speed attritor (bead mill) led to the result that, within the accuracy of the measurements, all of the mechanical power put into the system was dissipated to Joule's heat [12, 13]. In this sense, bead mills are "energy killers". The energy is transferred to the mill bases, causing them to gain heat which, as a result, necessitates the consumption of cooling water. That is why it may make economic sense to utilize the generated heat. The energy dissipated from 20 pieces of dispersing machinery is sufficient to heat 4500 m² plant area the whole year around even with air ventilation running parallel [14].

A handicap of bead mills is their relatively high loss of mill base yield since some of the material remains in the milling chamber that cannot be recovered. A further drawback is the necessity of cleaning the milling chamber together with the milling bead filling if the product is changed. These downsides are minimized in the so-called basket mills. Basket mills consist of baskets in which the milling beads are placed. The principle features of basket mills and their mode of operation are shown in Figure 4.14. The basket holding the beads is lowered into a pre-dispersed mill base (Figure 4.14-1). Then, the shaft is started. It propels a pump for the mill base while keeping the milling beads within the basket in motion. The basket itself remains stagnant. To facilitate mixing, an impeller blade may be connected to the tip of the shaft. Due to the circulation of the mill base, it continuously passes through the milling basket where dispersion takes place. The beads are kept from leaving the basket for example by a sieve at its bottom (Figure 4.14-2).

When the dispersion is completed, the basket is lifted out of the mill base and the material within the basket is thrown out by letting the shaft run for a short period of time (Figure 4.14-3). The basket may then be cleaned in a separate step (Figure 4.14-4). Of course, basket mills are operated only discontinuously so that batch sizes of a few tons at most are fabricable. On the other hand, if larger batch sizes are to be produced, the use of conventional bead mills again becomes preferential since the fraction of mill base that cannot be recovered is lower again.

4.5 Determining dispersion time

It is easily comprehended that the duration of dispersing has an influence on the success of dispersing. In the case of discontinuously operated dispersion devices in which the mill base is completed emptied into the vessel where the dispersion is performed, the time of dispersion t is identical to the running time t_{run} of the apparatus. It can be measured simply with a stop watch.

In discontinuously operated dispersion media such as bead mills, matters are somewhat more complicated. As already pointed out, the mean dwell time calculated from Equation 4.4, which is the mean time of dispersion for one single pass through the machine, is just a mean value. Depending upon the intensity of the axial cross mixing in the milling chamber, the dwell time is a distributed quantity. On the other hand, two modes of operation in continuous dispersion, the single pass mode and the recirculation mode, must be distinguished from one another.

In the single pass made (Figure 4.15) the mill base is pumped from a feed container to the milling chamber.

The mill base volume V_{MB} passes through the milling chamber and is discharged into a collecting vessel. In this case, the mean dispersion time is represented by Equation 4.4. When n consecutive passes are run, the mean time of dispersion is

Equation 4.5 $$t_m = n \cdot \frac{V_{free}}{\dot{V}_{MB}}$$

whereby it should be kept in mind that a larger number of passes not only lengthens the duration of the dispersion but also has advantages concerning the dwell time distribution. So, as far as the uniformity of dispersion time is concerned, it is better to make a larger number of quick passes instead of one lengthy pass. The drawback of the single pass mode is that it consumes more labour costs because the mills have to be turned off and on for each new pass and the fittings have to be changed.

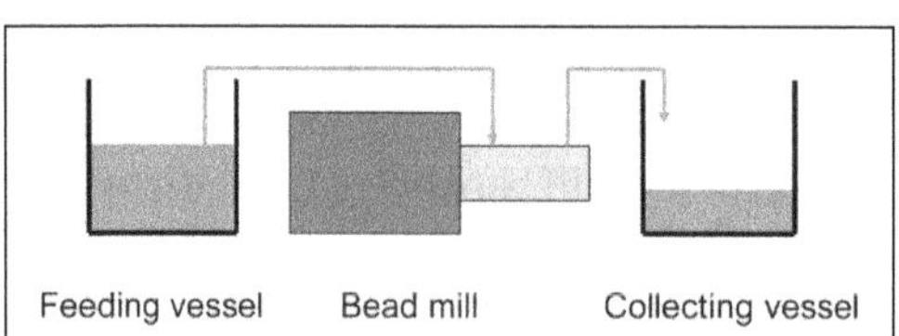

Figure 4.15: Operating a bead mill in the single pass mode

In the recirculation mode (Figure 4.16), the feeding and the discharge container are identical. The mill base is passed through the mill and is recirculated to the feed vessel. Sometimes the mill base in the container is stirred.

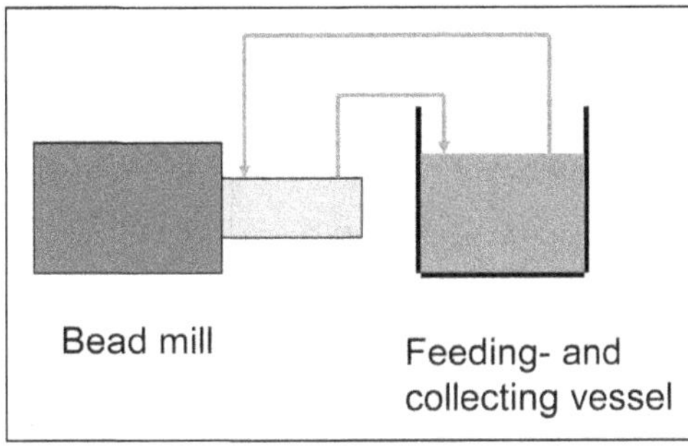

Figure 4.16: Operation of a bead mill in recirculation mode

No intervention what so ever is necessary while the dispersion is taking place. This facilitation is, however, relativized as the process control is lower and the dispersion takes longer for that reason. Even when the machines are run for a very long time, it is possible that a part of the mill base is not dispersed at all. In order to estimate the mean dispersing time in the recirculation mode, it is therefore mandatory to put some thought into the theoretical number of passes $n_{theor.}$ in the recirculation mode. This number states how many times the mill base volume V_{MB} with a volume flow of $\dot{V}_{MB}$ should have passed through the milling chamber during the operation time t_{run} of the mill. Further considerations lead to

Equation 4.6 $$n_{theor.} = \frac{\dot{V}_{MB}}{V_{MB}} \cdot t_{run}$$

If the theoretical number of passes is lower than one, then the dispersing time was not sufficient to send all of the mill base at least once through the milling chamber. In dependency of the mixing intensity of dispersed and not dispersed mill base in the feed container, one may estimate after how many theoretical passes all of the mill base has passed the milling chamber at least once. The better the mill base is agitated in the feed vessel, the more theoretical passes are necessary.

The mean dispersing time is calculated in the recirculation mode from the free volume in the milling chamber, the volume of mill base being dispersed and the run time of the machine. Since only the fraction V_{free}/V_{MB} is able to be in the milling chamber at once, the mean dispersing time is found to be

Equation 4.7 $$t_m = \frac{V_{free}}{V_{MB}} \cdot t_{run}$$

A practical example demonstrates the usefulness of these thoughts. A nano-scaled silica was supposed to be dispersed as well as possible in a resin solution in order to obtain a transparent coating. This was done with a bead mill that had a milling chamber of 4 litres volume. The milling chamber was filled with milling beads to a bulk volume of 85 % so that a free volume of 2.3 litres was obtained. 100 litres of mill base were dispersed for a period of eight hours in the recirculation mode,

Table 4.2: Determination of the (mean) dispersing time for discontinuous dispersions and for dispersions in the single pass and recirculation modes, respectively.

Mode of operation	Number of passes	(mean) Dispersion time
Discontinuous	I	$t = t_{run}$
Continuous in pass mode	n	$\bar{t} = n \cdot \frac{V_{free}}{\dot{V}_{MB}}$
Continuous in recirculation mode	$\frac{\dot{V}_{MB}}{V_{MB}} \cdot t_{run}$	$\bar{t} = n \cdot \frac{V_{free}}{\dot{V}_{MB}} \cdot t_{run}$

whereby the mill base was pumped through the mill at a rate of 300 litres per hour. According to Equation 4.6, the number of theoretical passes was 300/100 · 8 = 24. So, one may assume that all parts of the mill base had passed the milling chamber a number of times. Under this condition, it makes sense to go on and calculate the mean dispersing time with Equation 4.7. This yields 8 · 2.3/100 = 0.184 hours or approximately 11 minutes which is, when compared to common discontinuous dispersions and in light of the apprehended dispersion result, not too long a time. Accordingly, the achieved dispersion result was insufficient. An estimation of these parameters before commencing the dispersion would probably have resulted in the selection of a longer machine run time to start with.

Table 4.2 presents an overview of equations used to calculate the number of passes and the mean dwell times for the different modes of operation of dispersion equipment.

4.6 Literature

[1] B. Schwegmann, Farbe und Lack 80 (1974) 311

[2] B. Schwegmann, T. C. Patton, Journal of Paint Technology 42, No. 550 (1970) 626

[3] H. E. Weisberg, Official Digest 36, No. 478 (1964) 1261

[4] F. K. Danial, Journal of Paint Technology 38, No. 500 (1966) 534

[5] R. I. Ensminger, Official Digest 35, No. 456 (1963) 71

[6] W. Bueche, Dissertation, "Der Zerkleinerungsvorgang auf Reibwalzenstühlen", Karlsruhe (1932)

[7] H. Potente, K. Kretschmer, J. Flecke, Polymer Engineering Science, 42, No. 1 (2002) 19–32

[8] A. Kamptner, P. Koch, Aufbereitungs-Technik 32 (1991) 159–164

[9] K. Schönert, K. Steier, Chem.-Ing.-Techn. 43 (1971) 773–777

[10] S. Bernotat, K, Schönert, "Size Reduction", Ullmann's Encyclopedia of Industrial Chemistry, VCH, Weinheim, 5th Ed., 1988, Vol. B2, Chapter 5

[11] H.-J. Henzler, "Continuous Mixing of Fluids", Ullmann's Encyclopedia of Industrial Chemistry, VCH, Weinheim, 5th Ed., 1988, Vol. B4, p. 576-577

[12] J. Winkler, farbe+lack 90 (1984) 244–250

[13] K. Engels, farbe + lack 71 (1965) 464

[14] K. Halsch, Journal für Oberflächentechnik 20 (1980) 486

5 Mechanical breakage of agglomerates

The breakage of agglomerates is the basic requirement of dispersions. In spite of the fundamental importance of this step, it is often the least understood. Still, there is a simple model which describes the conditions that must be met for the mechanical breakage of the agglomerates to take place. This model is presented in Chapter 5 in a qualitative and quantitative manner. Furthermore, the unique aspects of running so called "bead mills", which are of great industrial importance, are covered. The special case of dispersing nano-scaled pigments will also be discussed.

5.1 Measuring dispersion success

If the status of agglomerate fracturing during a dispersion is to be studied, it is mandatory that the degree of deagglomeration is brought into connection with a measurable mill base or paint property. Particle size measurements fail in this respect, at least in the case of paint formulations, because of technical difficulties. The determination of particle sizes from electron microscopic pictures of cross sections of cured paint films is very tedious [1]. Contrarily, colorimetric measurements such as the determination of colour strength according to Kubelka-Munk or of the scattering power have proven to be useful.

The so called colour strength of a substance or an optically completely hiding paint film is calculated from the remission β_∞ of the colour filter (red, green or blue) with which the highest absorption is measured. The colour strength is the quotient between the absorption coefficient K and the scattering coefficient S of the body or paint film. Equation 5.1 is called the Kubelka Munk function.

Equation 5.1 $$F = \frac{K}{S} = \frac{(1-\beta_\infty)^2}{2\beta_\infty}$$

β_∞ can only obtain values between zero (0 % remission) or unity (corresponding to 100 % remission). Figure 5.1 shows the colour strength function K/S in dependency of the remission β_∞. Since it is customary to measure the colour strength in a remission interval between 0.25 and 0.8, the graph is plotted only in this range. When moving to smaller remission values, the Kubelka-Munk function is very steep.

Jochen Winkler: Dispersing Pigments and Fillers

ISBN 978-3-86630-852-7

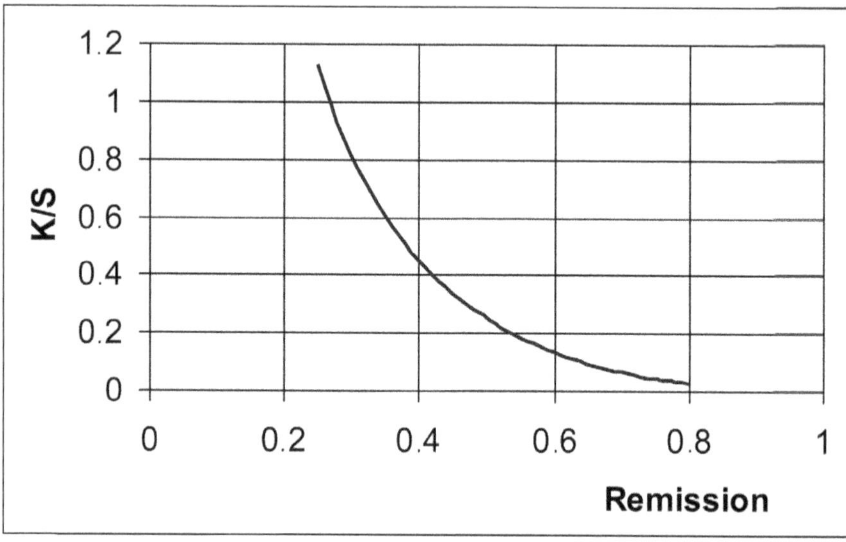

Figure 5.1: The Kubelka-Munk function (Equation 5.1)

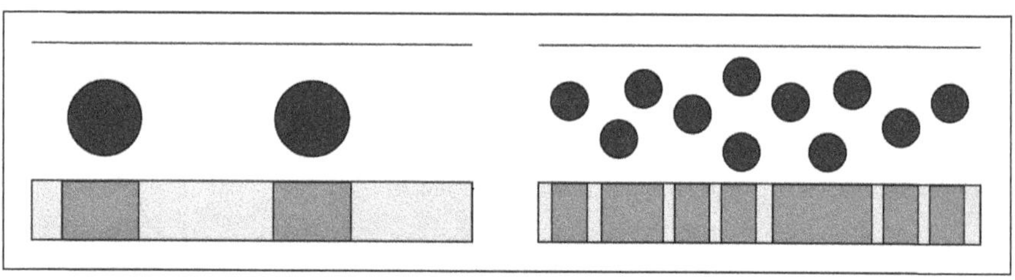

Figure 5.2: Comparison of surface coverage of different particle sizes

Measuring the colour strength as an indicator for dispersion results is particularly convenient when coloured pigments are dispersed along with titanium dioxide. In that case, the contribution of the absorption coefficient to the colour strength comes from the coloured pigment alone, whereas the scattering coefficient may be attributed to the distribution of the white pigment. The absorption coefficient increases when the particle distribution (degree of dispersion) of the coloured pigment improves. The reason for that is that the pigment particles are more evenly spaced and therefore can absorb light more efficiently. This is shown schematically in Figure 5.2. The eleven small (primary) particles on the right side of Figure 5.2 have the same volume as the two (agglomerates) on the left side because their diameters are only 57 % of those of the larger agglomerates. More of the light falling in from the top is therefore absorbed by them, as indicated by

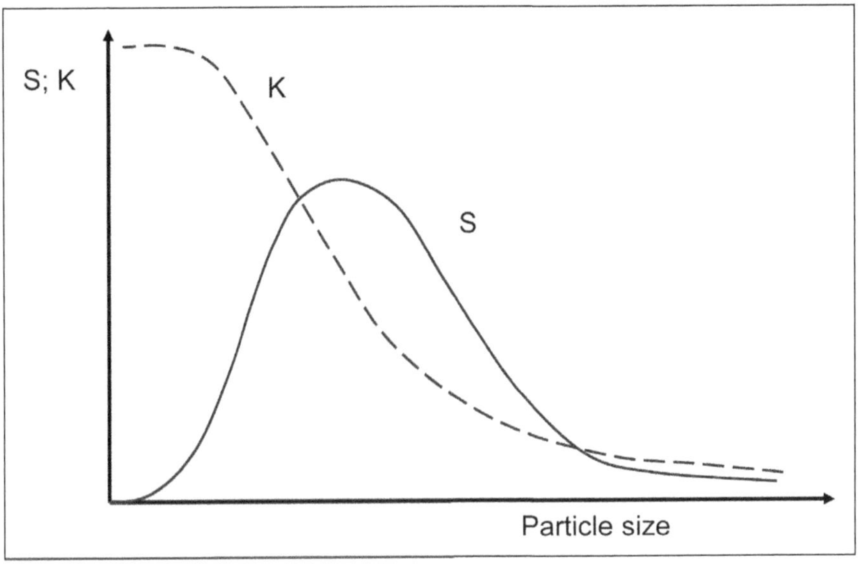

Figure 5.3: Dependency of the absorption coefficient K and the scattering coefficient S on pigment particle size (schematic drawing)

their "shadows" in Figure 5.2. Figure 5.3 shows the dependency of the absorption coefficient from the particle size in a schematic way. In Figure 5.3, agglomerates would act like large particles.

The scattering coefficient also depends upon the particle size and runs through a maximum at a particle diameter which is approximately half the size of the wavelength of the incident light [2] (see Figure 5.3).

A further property that often depends upon the particle distribution achieved while dispersing is the gloss of a coating. Devices for measuring gloss are described in the standards DIN 67 530, ASTM D 523-78 and ISO 2813, for example. A circular light beam of defined cross section is directed to the surface of a substrate at a defined incident angle α. The intensity of the reflected beam is determined with the "reflectometer" (DIN 67 530) at the emergent angle using different apertures[1]. Figure 5.4a shows the measuring principle on an ideally plane (specular) surface. If the surface is structured, the light is not only reflected in the emergent angle, but to some degree diffusely also (Figure 5.4b).

Accordingly, a part of the reflected light is blended out by the aperture so a lower gloss is measured. The standard DIN 67 530 distinguishes between high gloss,

[1] An aperture is an opening (blend) that makes a light ray more collimated

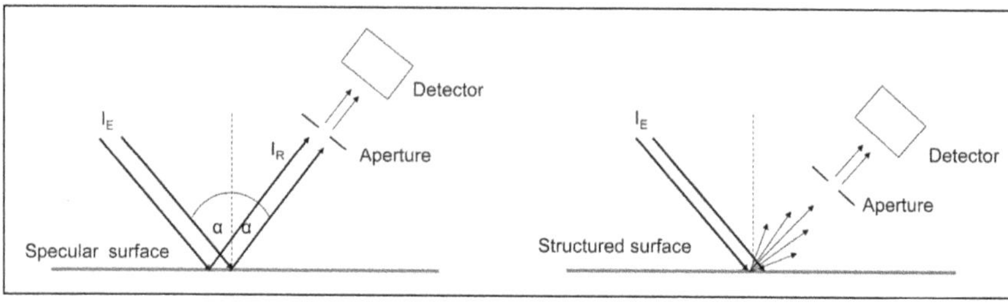

Figure 5.4: a) Determination of gloss of a specular surface
b) Determination of gloss of a non-specular, rough surface

medium gloss and matt gloss that are determined at 20°, 60° and 85° incident angle, respectively. (For a given surface, the gloss increases with increasing incident angles.) Poorly dispersed pigments have both more and larger agglomerates that structure the paint surface and therefore reduce the gloss. If the gloss is to be taken as a measure for the achieved dispersion result, care must be taken that the film thicknesses of the coatings are identical in each case. Higher film thicknesses lead to higher gloss values and vice versa. Gloss measurements are particularly suitable for monitoring the dispersion state of physically drying paints or mill bases with a high pigment loadings.

The viscosity of mill bases also depends upon the pigment particle distribution. With increasing degree of dispersion, the resistance to flow due to the particles decreases so that the viscosity becomes less. At high loads, though, the viscosity may increase again (see Chapter 3.5.1). Viscosities of mill bases with pronounced flow anomalies (pseudoplasticity, dilatancy, thixotropy or rheopexy) are difficult to measure since their flow characteristics are highly dependent upon the handling of the sample, i.e. storage time and temperature.

Although not very accurate, the measurement of the fineness of grind with a Hegman gauge (or grindometer) according to ASTM D333, ASTM D1210, ASTM D1316 or DIN EN ISO 1524 is routinely done. The Hegman gauge consists of a steel block with a shallow, tapered groove cut into it. At the side of the block, a scale indicates the depth of the groove. A drop of mill base or paint is deposited at the deepest part of the groove and is then drawn out to the other, shallow side with a scraper. When viewed from the side at a slight angle, the place where pigment particles protrude through the surface of the film becomes visible and a reading can be taken from the scale. This value represents the "fineness of grind", "Hegman" or "grindometer" reading. When doing comparative measurements, the samples should have identical compositions in order for the measurements to really be comparable. When the samples are thinned with solvents, for example, lower finenesses, that is smaller particle sizes tend to be found. Finally, the measurements should always be carried out by the same person because of the inherent subjective error. The DIN EN ISO 1524 standard states that one and the same person has 95 % probability to lay within 10 % error in consecutive readings. For two separate people, the deviation doubles to +/-20 %.

When dispersing, all of these properties can develop differently. Under the condition of principle applicability, gloss and colour strength developments correlate agreeably. This can be attributed to the fact that on one hand, the absorption coefficients of coloured pigments do not change much once the pigment particles have reached a certain, small size (Figure 5.3) while on the other hand, from a certain fineness onwards, the particles are too small to roughen the surface of a coating so that the gloss does not change any more. On the contrary, viscosities can still change when colour strength and gloss have reached final values. Hegman gauge readings are only crude indicators for the achieved dispersion state and their measurement is imprecise.

In practice, when wanting to determine the achieved dispersion state, the best choice will be a method that both gives a reliable measure for the quality of the dispersion and which is advantageous in view of the effort necessary to obtain a value. This will be different for construction paints as opposed to, for example, automobile paints or pigment pastes used for tinting. The latter can be formulated using very costly pigments so that optimal dispersion is also of high economic relevance. Finally, in the case of tinting pastes, technical issues dictate the necessity for optimal and reproducible dispersions, for example if they are used in automatic colour mixing systems where different colour strengths from batch to batch would inevitably lead to varying colours of the mixed paints.

5.2 Principle of mechanical breakage; hammer–walnut experiment

An experiment as shown in Figure 5.5 is extremely helpful for understanding the principles of the mechanical breakage of agglomerates in dispersing equipment. A titanium dioxide white pigment is thoroughly dispersed in a resin solution with a bead mill so that it is viable to assume that the pigment is dispersed completely.

The pre-dispersed white paste is placed into the milling vessel of a high speed impeller and dry coloured pigment powder is added while mildly stirring with the impeller blade. Then the dispersion is started at a constant rotational speed of the impeller shaft. The ratio of coloured pigment to titanium dioxide should be selected in such a way that the colour strength measurements can be performed without any further additions. This means that the remission of the mill base should lay somewhere between 0.25 and 0.8, depending upon the dispersion state. Colour strength is best determined on the wet paint. Naturally, the pigments must

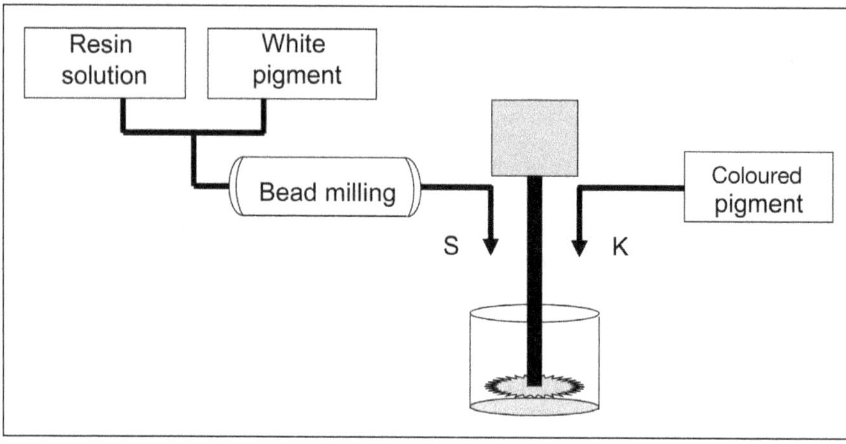

Figure 5.5: Experiment for investigating the influence of time and rotational speed of the shaft of a high speed impeller separately

be stable against flocculation in the mill base. It may therefore be necessary that suitable additives are found in preceding tests.

While the high speed impeller is running, small samples of mill base are taken from time to time and colour strength measurements of them are made. When the colour strengths no longer develop, the dispersion is stopped and a fresh, identical batch of mill base is fed into the grinding vessel and the same experiment is repeated, yet with another rotational speed.

Since the white pigment was already dispersed prior to the experiment, changes in colour strength are caused only from the dispersion of the coloured pigment. Figure 5.6 shows the result of such a series of dispersions with colour strength plotted against the dispersion time at different rotational speeds. At every rotational speed, the colour strengths first develops rapidly and then less quickly until they plateau into the final value. There are characteristic final colour strengths for every chosen rotational speed choosen, whereby these are higher for higher rotational speeds. The colour strength development curves do not intersect. Higher rotational speeds not only increase the final colour strength, but also lessen the time needed to reach the final values.

Only one possible interpretation for this experimental result is feasible. For the agglomerates to be dispersed, they must be transported into places where they can be mechanically stressed. If they are dispersed when they get into these locations or not depends upon their strength. The experiment shows that it is only a question of time for all the agglomerates to pass through the sites where the agglomerates are stressed the most and that there are agglomerates that are so

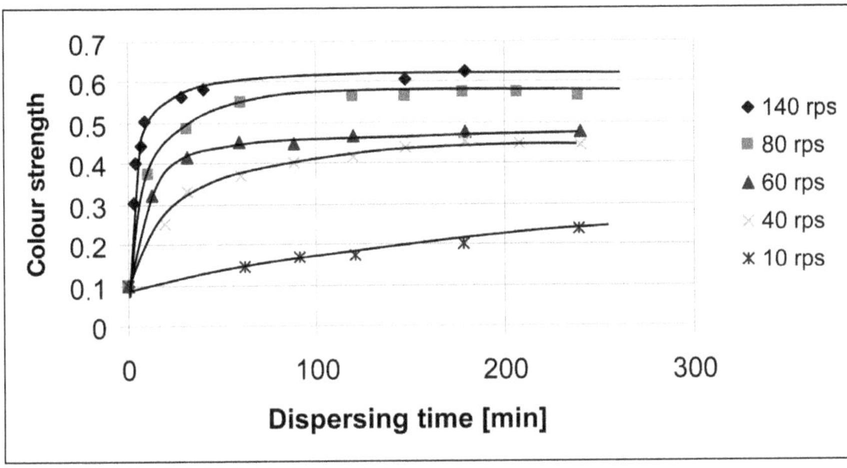

Figure 5.6: Time dependent colour strength developments for dispersing a coloured pigment in a white paste at different rotational speeds (rps=revolutions per second)

strong that they are not disrupted there. Apparently, agglomerates that do not get dispersed at a certain rotational speed certainly may be dispersed when the shaft runs at a higher speed.

In a high speed impeller, the pigments are fractured by shearing. Initially, there are many places in the milling vessel where the shear is high enough to cause the agglomerates to break. That is why the colour develops more rapidly at the beginning of the dispersion. With increasing duration, the agglomerates which have a lower strength deplete in the mill bases and the sites where there is sufficient shear for further dispersion to take place become less and less. Supposedly, the highest exposure is at the tip of the impeller blade since that is where the highest shear rate[2] prevails. It takes a certain amount of time until every agglomerate has passed through the sites with the highest shear conditions. Apparently, the agglomerates that weren't dispersed there remain unharmed even after a second, a third or any further pass.

By increasing the rotational speed, both the transportation rate of the agglomerates and the mean as well as the maximum shear stress is increased. This results in a faster development of the colour strength and, on the other hand, to a higher final value.

An ostensive picture for dispersing agglomerates could be the following: Walnuts (agglomerates) are thrown on a table (milling chamber) and a hammer (milling beads in motion) is used to randomly (eyes closed) hit the surface of the table. In

[2] The shear rate ist the gradient of the flow velocity within a liquid. Between to sites Between to sites in a liquid at a distance a with a difference in flow velocity of Δv the shear rate is $\Delta v/a$. See Chapter 3.5.

the course of time, every walnut will have been struck at least once. Unlike the walnuts with a strong shell, those with a softer shell will be cracked (dispersed). In order to crack the tough walnuts too, the hammer has to hit with greater force.

The hammer walnut analogon is a formidable description for the use of any type of dispersion or grinding machinery. In any kind of comminution process whatsoever, the particles have to be stressed and the stressing must be strong enough for the agglomerates (when dispersing) or the particles (when grinding) to break. When all other parameters are fixed, the probability for particles to encounter a stress situation depends only upon the duration of the comminution process, whereas the probability for a size reduction to occur depends upon the relationship between the intensity of the stress and the ability of the particle or the agglomerate to withstand it. This beautifully simple model is therefore sufficient to study, understand and optimize any dispersing or grinding process.

5.3 Dispersion equation

Provided that the probability for the agglomerates to be stressed and the probability for them to break can be put into mathematical terms, then, following general rules of probability calculus, the overall probability for the mechanical breakage to take place may be obtained as the product of the two partial probabilities. Chapter 5.3 shows that this is indeed possible.

5.3.1 Stress probability

The model that leads to an expression for the time dependent stress probability p_t is shown in Figure 5.7.

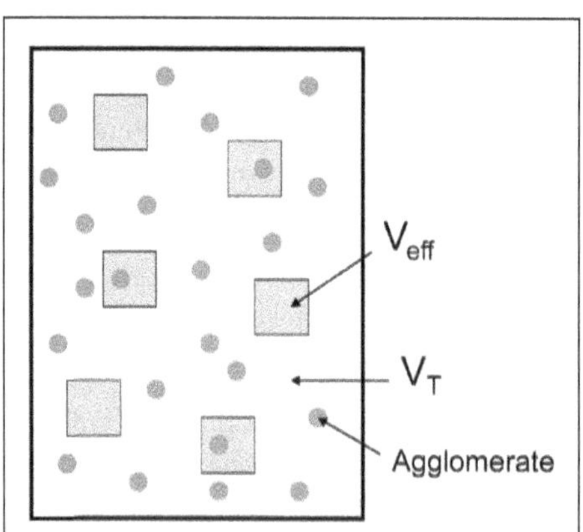

Figure 5.7: Drawing to illustrate the derivation of an expression for the stress probability

Figure 5.7 shows the drawing of, for example, a milling chamber of a bead mill with a total volume V_T. Within the chamber there are certain volume elements in which agglomerates are stressed. Put together, all of these sites add up to the effective volume V_{eff}. If the agglomerates are distributed evenly within the chamber, then their number multiplied by the quotient V_{eff}/V_T yields the number of agglomerates that are within the effective volume elements at any time. To come to an expression for the stress probability one may formulate that the decrease in the number of non-stressed agglomerates, that is those that have not yet been in an effective volume element, is proportional to the number of agglomerates in the effective volume elements at any time.

Equation 5.2 $$-\frac{dN_n}{dt} = k \cdot N_n \cdot \frac{V_{eff}}{V_T}$$

with

$-\frac{dN_n}{dt}$ = decrease in the number of pigment agglomerates not yet stressed
N_n = number of non-stressed pigment agglomerates
k = proportionality constant with the dimension s^{-1}

Rearranging of Equation 5.2 followed by integrating between the limits $t=0$ und t leads to Equation 5.3.

Equation 5.3 $$ln \frac{N_{n,t}}{N_{n,t=0}} = -k \cdot \frac{V_{eff}}{V_T} t$$

$N_{n,t}$ depicts the number of agglomerates that have not been in an effective volume element at the time t, whereas $N_{n,t=0}$ stands for the number of agglomerates initially present (at $t=0$). Eliminating the logarithm in Equation 5.3 leads to Equation 5.4.

Equation 5.4 $$\frac{N_{n,t}}{N_{n,t=0}} = exp\left(-k \cdot \frac{V_{eff}}{V_T} \cdot t\right)$$

At any time during a dispersion, the sum of the stressed $N_{d,t}$ and non-stressed agglomerates $N_{n,t}$ is identical to the number of agglomerates initially present.

Equation 5.5 $$N_{n,t=0} = N_{n,t} + N_{d,t}$$

Inserting $N_{n,t}$ from Equation 5.5 into Equation 5.4 leads to an expression for the stress probability p_t.

Equation 5.6 $$p_t = \frac{N_{d,t}}{N_{n,t=0}} = 1 - e^{-k \frac{V_{eff}}{V_T} t}$$

Because of the large number of agglomerates, the fraction of stressed agglomerates divided by the number of agglomerates initially present is identical to the probability for a stress situation to take place[3].

A comparison of the physical dimensions of the left and right hand side of Equation 5.2 shows that the proportionality constant k has the dimension of a reciprocal

[3] If, for example, at the time t 200 of the 1000 agglomerates initially present have been in effective volume elements, then the stress probability would be 200/1000 = 0.2, or 20 %.

time. k is a rate constant and $k \cdot V_{eff}$ therefore reflects the effective volume per time unit in a piece of dispersion machinery. k gets larger, so the dispersion gets faster, the faster the agglomerates are transported into effective volume elements, that is, the more the mill base is kept in motion and the more evenly the effective volume elements are distributed.

5.3.2 Breaking probability

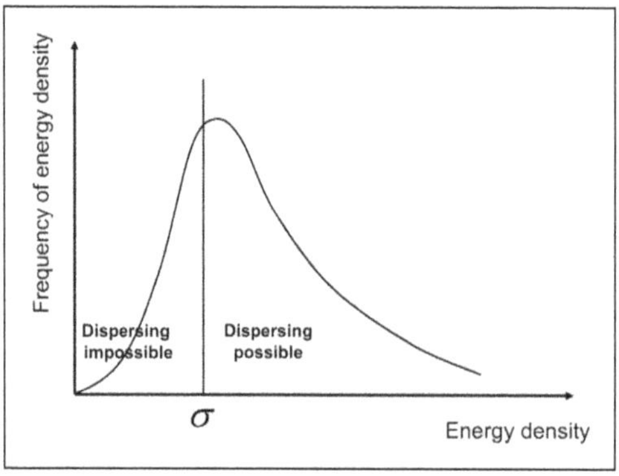

Figure 5.8: Schematic representation of the energy density distribution in a dispersion device in relation to the agglomerate strength

It was shown in Chapter 2.5 that the strengths of agglomerates influence their ease of dispersion (compare Figure 2.10). Softer agglomerates are easier to disperse than ones that are tougher. At first approach, one may assume that the breaking probability is inversely proportional to agglomerate strength (see Equations 4.1 and 4.2). The agglomerate strength has the physical dimension of *work* per *volume*. For principle reasons, whatever causes agglomerates to break must have the same physical dimension, namely of an energy density. If it is taken for granted that the deagglomeration takes place in a shear field, then the energy involved should be a kinetic energy. The kinetic energy in a dispersion device is certainly a distributed quantity. There are places with slower and faster flow velocities. This is schematically shown in Figure 5.8. An agglomerate with a strength σ can only be dispersed in a site where the energy density is larger than its strength. Naturally, the efficiency of the transfer of energy density onto the agglomerate will also play a role.

Since the energy introduced into the dispersion can only operate within the mill base phase, that is within V_T, one can write that the decrease in the number of non-dispersed agglomerates N_n with increasing energy density E/V_T is proportional to the number of agglomerates present and inversely proportional to their strength σ.

Equation 5.7
$$-\frac{dN_n}{d(E/V_T)} = a \cdot \frac{1}{\sigma} \cdot N_n$$

Equation 5.7 shows that the proportionality constant a is dimensionless. It has the meaning of a transfer constant and is a measure for the efficiency with which the energy density is used to disperse the agglomerates. Integration of Equation 5.7 within the limits $E/V_T = 0$ to E/V_T leads to

Equation 5.8 $$\frac{N_{n,\varepsilon}}{N_{n,\varepsilon=0}} = exp\left(-\frac{a \cdot E}{\sigma \cdot V_T}\right)$$

Once again, the number of agglomerates initially present $N_{n,\varepsilon=0}$ is the sum of those which are dispersed $N_{d,\varepsilon}$ and those which aren't $N_{n,\varepsilon}$.

Equation 5.9 $$N_{n,\varepsilon=0} = N_{n,\varepsilon} + N_{d,\varepsilon}$$

Inserting $N_{n,\varepsilon}$ from Equation 5.9 into Equation 5.8 leads to an expression for the breaking probability p_ε.

Equation 5.10 $$P = \frac{N_{d,\varepsilon}}{N_{n,\varepsilon=0}} = 1 - e^{-\frac{aE}{\sigma V_T}}$$

5.3.3 Total probability

It should be kept in mind that the two partial probability terms are completely independent from one another. p_t expresses the probability with which the agglomerates are stressed in a dispersion device. It depends upon the distance of separation between the effective volume elements and on the rate at which the agglomerates are brought into them as well as on the duration of the dispersion (see Equation 5.6). p_ε expresses the probability with which they are dispersed when they are stressed. It depends upon the relationship between the energy density and agglomerate strength as well as on a constant which characterises the efficiency of the energy density transfer to the agglomerates (see Equation 5.10).

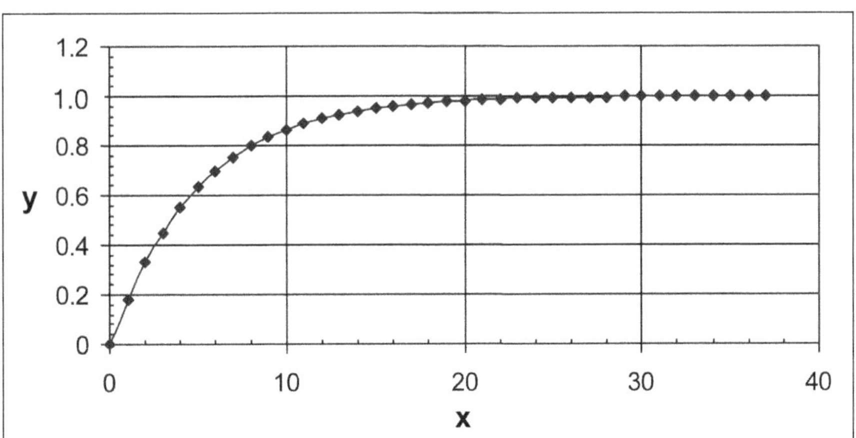

Figure 5.9: Function $y = 1 - e^{-const\,x}$. For "const.", a value of 0.25 was used in the calculation

For the agglomerates to be dispersed, both conditions must be met simultaneously, i.e. the agglomerates must be stressed and must be stressed hard enough (hammer-walnut analogon, see Chapter 5.2). Under these circumstances, the overall (total) probability p_T is the product of the two partial probabilities (dispersion Equation).

Equation 5.11 $$p_T = p_t \cdot p_\varepsilon = \frac{N_d}{N_{n,t=0}} = \left(1 - e^{-\frac{kV_{eff}}{V_T} \cdot t}\right) \cdot \left(1 - e^{-\frac{a}{\sigma} \cdot \frac{E}{V_T}}\right)$$

whereby N_d depicts the number of dispersed agglomerates.

The two single probabilities in Equation 5.11 are of the type

Equation 5.12 $$y = 1 - e^{-const \cdot x}$$

This function is displayed in Figure 5.9. With increasing x values, y first develops quickly and then slower until y asymptotically approaches the value $y = 1$.

In Figure 5.10, the total probability p_T of Equation 5.11 is plotted against the time of dispersion for different breaking probabilities or energy densities, respectively, whereby $k \cdot V_{eff}/V_T$ was taken to be 0.2 s^{-1}. Figure 5.10 reproduces the experimental conditions of Chapter 5.2 (see Figure 5.6). Figures 5.10 and 5.6 become comparable under the assumption that the colour strength (y-axis in Figure 5.6) in the dispersion experiment is directly proportional to the fraction of dispersed agglomerates (y-axis in Figure 5.9). The principle consistency in the development of the two curves indicates that the probability equation (dispersion equation) and therewith the model of thought that led to it, is well able to describe reality. A very important finding is that a lack of energy density cannot be compensated by dispersing for a longer period of time!

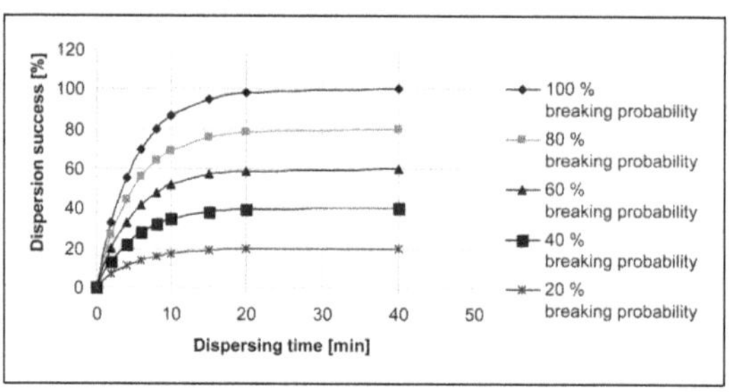

Figure 5.10: Plot of time dependent dispersion success at different breaking probabilities

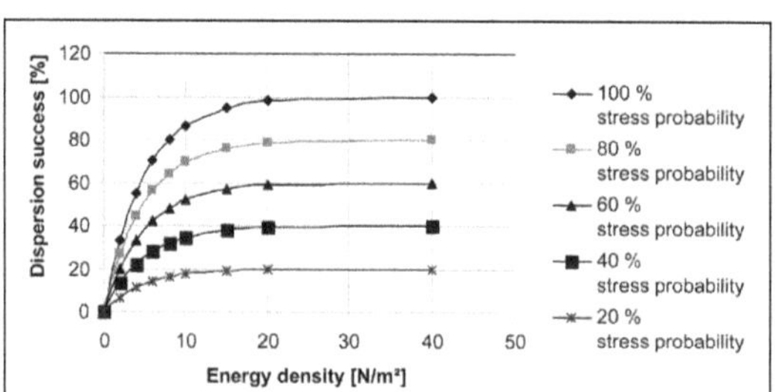

Figure 5.11: Plot of energy density dependent dispersion success at different stress probabilities

Figure 5.11 marks the opposite case where dispersion success is plotted as a function of the energy density employed during the dispersion for different stress probabilities. This graph stipulates that a lack of dispersion time may not be compensated by introducing higher energy densities.

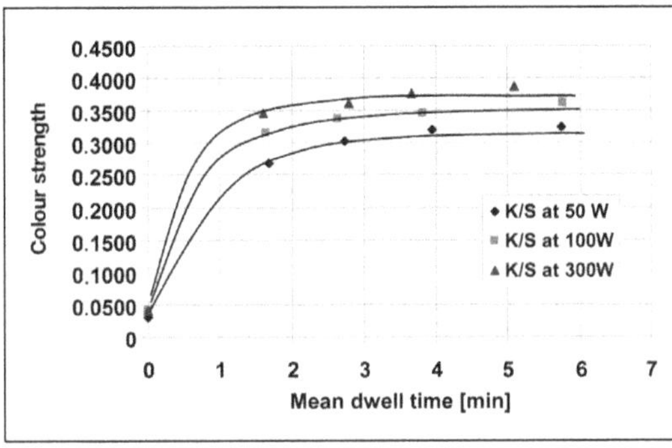

Figure 5.12: Colour strength development in dependency of the mean dwell time (single pass mode in a bead mill) at mechanical powers of 50 W, 100 W and 300 W

Figures 5.12 and 5.13 demonstrate a dispersion of a copper phthalocyanine pigment in a pre-dispersed white paste in a laboratory bead mill. In Figure 5.12 the colour strength development is plotted against the dispersion time at different mechanical powers P [4] of the shaft. This experiment depicts the situation of Figure 5.10. In contrast to that, Figure 5.13 denotes the situation as shown in Figure 5.11, the development of colour strength as a function of the mechanical stirring power at different durations of dispersing (mean dwell times of the mill bases in a bead mill). So, the behaviour that is expected from Equation 5.11 is confirmed completely in real dispersing experiments.

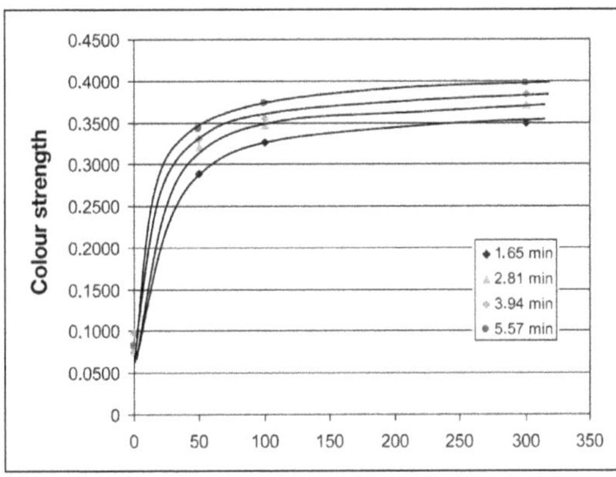

Figure 5.13: Colour strength development in dependency of the mechanical power (single pass mode in a bead mill) at mean dwell times of 1.65, 2.81, 3.94 and 5.57 minutes

5.3.4 Determination of the energy density

The dispersion Equation 5.11 has two independent variables, namely dispersing time and energy density.

In Chapter 4.5, it was already shown how the dispersing time can be estimated when operating dispersing machines in a continuous manner. However, nothing

[4] E/V_T is directly proportional to the mechanical power

as yet has been said about the energy density. When a dispersion is carried out, mechanical power P is transferred onto the mill base. Power has the physical dimension work per time, or joule/second = watt. In the case that the mechanical power is introduced by a rotating shaft, it can be calculated from the rotational speed n and the torque (momentum) M of the shaft.

Equation 5.13 $P = 2\pi \cdot n \cdot M$

The mechanical power of the shaft causes the temperature of the mill base rise. To make sure that the mill bases are not harmed by heat, the mills are therefore cooled with water. Independent of if the dispersion is run continuously or discontinuously, eventually a power equilibrium is reached at which neither the mill base temperature, nor the cooling water temperatures change. In this state of thermal equilibrium, just as many joules per time unit are introduced into the milling chamber as joules flow out again in the form of heat. That is why there are always just as many joules "active" in the dispersing machine. If, for example, a dispersion is carried out at a power of 1000 watt, then, at thermal equilibrium, 1000 joules are effective. It is this energy that must be divided by the total free volume V_T to come to the energy density which is part of Equation 5.11.

Since mill bases normally display temperature dependent flow properties, the mechanical power generally changes during the course of a dispersion. When dispersing in the single pass mode, Equation 5.11 can still be used to evaluate, for example, a colour strength development curve even though the mechanical power was higher when the dispersion was started. Keeping the flow rate of the mill base and the cooling water at a fixed level, one must only wait until the thermal equilibrium is reached before taking a sample. This sample is then characteristic for the conditions chosen, including the energy density. When dispersing discontinuously, as is often done in the quality control of pigments, then, due to the higher torque at lower temperatures, the mechanical power is higher at the beginning of the dispersion than when the power equilibrium is reached. Depending upon the extent of this affect and based on what conclusions are to be drawn from the experiment, one must consider if this influence can be tolerated or not. Otherwise, machines should to be used that allow the mechanical power at which the dispersion is supposed to run to be selected in advance. These machines constantly measure the torque and adjust the rotational speed according to Equation 5.13 in such a way that the preselected mechanical power is always maintained. But also for continuous dispersions these machines may be useful, especially if the results are to be exactly reproducible such as in quality control, since the cooling water can easily be ten degrees colder in winter than in summertime and this will cause the torque to be higher. Figures 5.14 and 5.15 compare the course of millbase temperature, torque, rotational speed and mechanical power while dispersing at constant rotational speed (Figure 5.14) and at constant mechanical power (Figure 5.15), respectively.

It is very important not to confuse the dissipated energy with the energy calculated from the product of the mechanical power and duration of the dispersion!

Equation 5.14 $E = P \cdot t$

The energy found from Equation 5.14 is not the energy that is assignable to the dispersion success. Otherwise it should be possible to obtain to excellent dispersion results just by stirring moderately, provided this is done long enough. Yet, the experimental results prove the opposite (Figures 5.6, 5.11 and 5.12): A lack of dispersing time may principally not be compensated by a higher mechanical power and vice versa, provided either the stress probability or the breaking probability have reached 100 % already. There are quite a few papers that describe dispersing tests in which the "dispersion energy" according to Equation 5.14 was changed and in which the degree of deagglomeration of pigments is assigned to that energy. A closer look reveals that, actually, only the duration of the dispersions were changed whereas the rotational speed, and therewith the mechanical power, was kept more or less constant. Naturally, from these experiments it is not possible to deduce that the dispersion energy of Equation 5.14 is causal for agglomerate breaking.

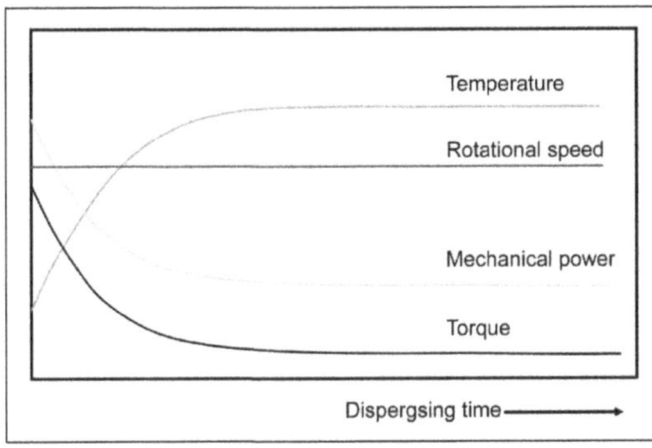

Figure 5.14: Change of mill base temperature, torque and mechanical power while dispersing at constant rotational speed

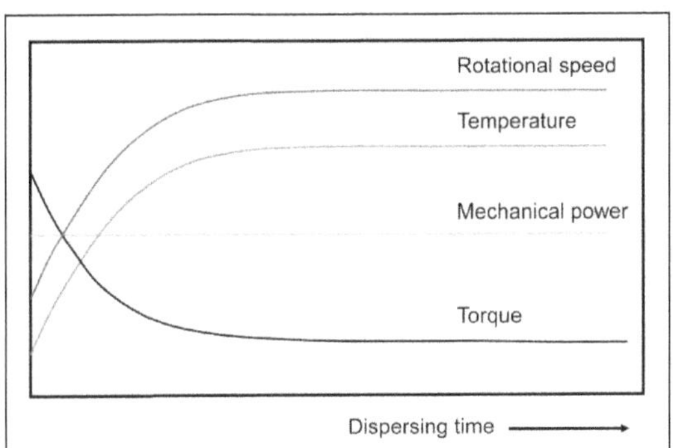

Figure 5.15: Change of mill base temperature, torque and rotational speed while dispersing at constant mechanical power

5.3.5 Colour strength development function

A time dependent colour strength development formula was proposed by von *Pigenot*[3]. Pigenot dispersed pigment pastes with a three roll mill and determined the colour strength as a function of the number of passes, calling that "energy". In reality, of course, not the energy, but the duration of the dispersion was changed. His colour strength development formula was based on a "second order kinetics" approach, whereby the decrease in the number of agglomerates is proportional to the square of the number on agglomerates present.

Equation 5.15 $$-\frac{dN_n}{dt} = k_h \cdot N_n^2$$

In Equation 5.15, k_h is a proportionality constant. Equation 5.15 leads to a "hyperbolic" type of colour development function of the form:

Equation 5.16 $$\frac{1}{F_t} = \frac{1}{F_E} + \frac{1}{k_h \cdot F_E^2} \cdot \frac{1}{t}$$

When plotting the reciprocal colour strength F_t against the reciprocal dispersion time t, a straight line is found from which the final colour strength F_E can be taken from the intercept with the y-axis whereas the proportionality constant is obtained from the slope. The colour strength of the mill base prior to dispersing F_S is set to zero.

Equation 5.15 implies that the agglomerates disperse each other. The fact that there is no experimental evidence for that presumption was criticized [4]. Schmitz and co-workers therefore suggested an alternative colour strength development function that is based on a "first order kinetics" [5]. The derivation of this formula starts, analogous to Equation 5.2, with the approach that the decrease in the number of agglomerates is proportional to their number at any time.

Equation 5.17 $$\frac{dN_n}{dt} = k_s \cdot N_n$$

Starting from Equation 5.17, the same mathematical steps that led to Equation 5.4 yields Equation 5.18.

Equation 5.18 $$\frac{dN_{n,t}}{N_{n,t=0}} = exp\,(-k_s \cdot t)$$

That the colour strength at the beginning of a dispersion is already larger than zero is acknowledged by considering a number of N_S agglomerates to be already broken before the start of the dispersion. At the end of the dispersion, when all agglomerates that were dispersible under the prevailing conditions are comminuted and the final colour strength is reached, N_E agglomerates are dispersed. $N_{n,t}$, the number of agglomerates that are not yet dispersed at the time t may be expressed by the difference between the dispersible agglomerates that were dispersed N_E and those that were finally dispersed at he time t $N_{d,t}$.

Equation 5.19 $$N_{n,t} = N_E - N_{d,t}$$

Furthermore, the number of agglomerates not dispersed at the beginning, $N_{n,t=0}$, is given by

Equation 5.20 $N_{n,t=0} = N_E - N_S$

Inserting Equations 5.19 and 5.20 into Equation 5.18 leads to

Equation 5.21 $N_{d,t} = N_E - (N_E - N_S) \cdot e^{-k_S \cdot t}$

Assuming the colour strength F (see Equation 5.1) to be directly proportional to the number of dispersed agglomerates, the colour strength development function of *Schmitz* is gained.

Equation 5.22 $F_t = F_E - (F_E - F_S) \cdot e^{-k_S \cdot t}$

By measuring colour strength as a function of dispersing time t, the final colour strength F_E, the starting colour strength F_S and the rate constant k_S can be determined. This is best done by curve fitting using the least error square method[5] [6]. Otherwise, it is also possible to rearrange Equation 5.22 into a form that allows a graphical solution.

Equation 5.23 $-\ln\left(1-\frac{F_t}{F_E}\right) = -\ln\left(1-\frac{F_S}{F_E}\right) + k_s \cdot t$

A plot of the left hand side of Equation 5.23 against the dispersing time yields a straight line with the slope k_S and an intercept with the y-axis at $-ln(1 - F_S/F_E)$. Somewhat unfavourable about this graphical method is, however, that the final colour strength F_E must be estimated from the colour strength data in order to be able to calculate the left hand side of the equation at all.

With some more ado it is possible to determine the final colour strength by differentiation of Equation 5.22 towards the time [7].

Equation 5.24 $\frac{dF_t}{dt} = k_s(F_E - F_S) \cdot e^{-k_S \cdot t}$

A linear equation is obtained from 5.24 by taking the logarithm.

Equation 5.25 $\ln\frac{\Delta F_t}{\Delta t} = \ln\{(F_E - F_S) \cdot k_s\} - k_s \cdot t$

A plot of the logarithm of the quotient of the colour strength divided by the time interval versus dispersing time yields a straight line from whose slope the rate constant k_S is obtained, whereas the intercept with the y-axis gives a value for the difference in colour strength at the end and at the beginning of the dispersion. Using this Figure, the final colour strength can be calculated form Equation 5.22.

A comparison of Equation 5.2 and 5.17 shows that the exponent of the stress probability function $k \cdot V_{eff}/V_T$ and k_S, the rate constant in the colour strength devel-

[5] Spreadsheet application software like excel allow such calculations („solver")

opment formula of Schmitz, are identical. This opens the possibility to access $k \cdot V_{eff}/V_T$ from time dependent colour strength development experiments. Since the total volume in a milling chamber V_T is easily calculated separately, $k \cdot V_{eff}$, the effective volume per time unit in the milling chamber, can be determined experimentally.

When colour strength is measured as a function of the mechanical power, a power dependent colour strength development function with an efficiency constant k_P can be written analogous to the time dependent development (Equation 5.22).

Equation 5.26 $$F_P = F_E - (F_E - F_S) \cdot e^{-k_P \cdot P}$$

In this case, $a/(\sigma \cdot V_T)$ and k_P are identical to one another so that the quotient of the efficiency constant a and the mean agglomerate strength can be determined.

Following the reasoning of Chapter 5.1, it makes sense to approach dispersions in a practical manner and to utilize whatever paint property is most easily and reproducibly measured or otherwise is most meaningful in terms of dispersion success when evaluating time or power dependent dispersion experiments.

5.3.6 Experimental results; using the dispersion equation

In this section, a few examples for the utilization of the dispersion equation for interpreting power as well as time dependent colour strength development curves are discussed.

5.3.6.1 Dispersion experiment; variation of the bead filling degree

In Chapter 4.4.1 it was stated that the degree of milling bead filling (bulk volume) should be adjusted between 80 % and 90 % of the free volume in the milling chamber in order to achieve satisfactory dispersing results while, at the same time, keeping the wear within tolerable limits. Figure 4.11 in that chapter shows power dependent colour strength development curves for dispersions in a 1.4 litre pin disc mill (PM STS 1, Drais Company, Mannheim, Germany) using different amounts of milling beads while keeping the mean dwell time at a constant value. Under these circumstances, the quotient between the efficiency constant and the agglomerate strength can be found from the rate constant k_S of the colour strength development formula of Schmitz (Equation 5.22) for every individual dispersion. Table 5.1 lists the results for the experiments behind Figue 4.11. The value of the energy transfer efficiency constant a increases with increasing bead amounts. The ratio a/σ and, since the mean agglomerate strength is constant, the energy transfer efficiency constant a doubles between 500 g (26 % bead filling) and 1860 g (98 % bead filling).

Hence, the energy efficiency becomes better with increasing milling bead amounts although, in absolute terms, it is still marginal. Assuming a mean agglomerate strength of 1 N/cm^2, the energy transfer efficiency constant ranges between 0.01 and 0.02. This means that merely 1 % to 2 % per cent of the mechanical power input is actually used for dispersing. A plot of the exponent of the breaking probability function $a/(V_T \cdot \sigma)$ against the free volume of the milling chamber of the 1.4 litre mill is presented in Figure 5.16. A straight line can be laid between the data points. It yields a value of zero for the transfer efficiency constant when no beads are used ($V_b = 0$). This is a clear proof that the colour strength development formula of Pigenot (Equation 5.16) is based on a false assumption. Agglomerates definitely do not mill each other, but are dispersed in the shear field between the milling beads instead.

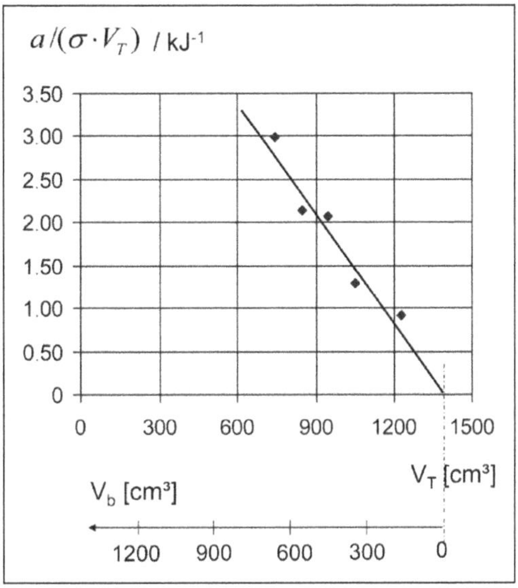

Figure 5.16: Dependency of the exponent of the breaking probability equation on the bead filling amount. V_b = bulk volume of milling beads.

Equation 5.11 shows that under these experimental conditions, where colour strength is determined as a function of the mechanical power, the final colour strength is determined by the stress probability. Just like the speed of deagglomeration, the stress probability increases with growing bead amounts. The rise in

Table 5.1: Colour strength development in dependency of the mechanical power at different milling bead amounts. Mean dwell time 16.9 minutes

Mass of 2 mm milling beads [g]	Final colour strength F_E	Milling bead volume V_b [cm³]	V_T=1400-V_b	$a/(V_T \cdot \sigma)$ in [kJ⁻¹]	a/σ [10^{-6} m²/N]
500	0.600	176.4	1223	0.923	1.13
1000	0.870	353.0	1047	1.285	1.35
1250	0.950	453.6	946	2.066	1.96
1575	1.050	555.8	844	2.143	1.81
1860	1.086	657.0	743	2.989	2.22

dispersion speed may formally be accounted to the fact that the effective volume elements are more closely spaced, so that the agglomerates are transported to them more quickly. Therefore, the agglomerates may be stressed more often. More evidence for the beneficial influence of a denser milling bead packing on dispersion results will be offered in Chapter 5.3.6.3.

5.3.6.2 Dispersion experiment; variation of dispersing time at different mechanical powers

Figure 5.6 in Chapter 5.2 shows the outcome of several discontinuous dispersion experiments with a high speed impeller equipment. The details of the experiments are given there. In this case, where the colour strength is determined in dependency of dispersion time for different mechanical powers, the constant of the colour strength development formula (Equation 5.22) depicts the exponent of the stress probability function (Equation 5.6) whereas the final colour strength depends upon the breaking probability or the mechanical power, respectively. The results are listed in Table 5.2.

The time dependent colour strength develops faster with increasing rotational speed of the shaft.

At rotational speeds between 10 s^{-1} to 140 s^{-1} the effective volumes increase from 0.01 cm^3 s^{-1} to 0.79 cm^3 s^{-1} because the mill base is circulated more rapidly so that the agglomerates reach the tip of the impeller blade more frequently. In this particular case, 0.79 cm^3 s^{-1} corresponds to about 0.4 % of the mill base volume. Comparable experiments with a bead mill, using 1 mm glass beads, gave effective volumes at a maximum of approximately one per cent of V_T.

Table 5.2: Evaluation of the dispersion from Figure 5.6

n [Hz]	P_{mech} [W]	F_E	$k*V_{eff}/V_T$ [1/h]	$k*V_{eff}$ [cm^3/s]
0	0	-	0	0
10.3	0.2	0.3655	0.167	0.01
20.5	4.2	0.4019	1.147	0.07
30.7	8.0	0.4278	1.855	0.11
40.5	14.0	0.4373	1.956	0.11
60.0	26.1	0.4486	5.557	0.32
70.4	31.0	0.5035	8.163	0.47
80.0	36.7	0.5415	7.090	0.41
90.0	43.6	0.5478	10.390	0.60
110.0	56.9	0.5570	11.670	0.67
140.0	78.8	0.5787	13.690	0.79

The final colour strengths develop from 0.3655 to 0.5787 due to higher breaking probabilities.

These and further examples for the quantitative use of the dispersion formula are published in [8].

5.3.6.3 Dispersing of nano-particles

Due to their small mass, nano-scaled pigments and fillers undergo only weak dispersive (van der Waals) interactions so that they tend to form very loosely packed agglomerates. This was already shown in Chapter 2.4. This leads to the low bulk densities typically found with this kind of material. Although agglomerates of low packing density are less well wetted (see Chapter 3.3), they are expected to be fairly easy to disperse because of the lower amount of contact sites between the particles and the lesser force of attraction at each contact site. Nevertheless, experience teaches that it is difficult to disperse nano-scaled pigments down to the primary particle range.

When using nano-scaled titanium dioxides as UV-absorbers in sun screens, they should ideally be transparent in them. Transparency becomes more pronounced, the better the agglomerates are dispersed. Bead mills are required to disperse nano-scaled particles. From experience, the transparency of nano-scaled titanium dioxides become better the smaller the milling beads are. Figure 5.17 shows the transmission of mill bases of a nano-scaled titanium dioxide UV-absorbers at a wave length of 520 nm, which is in the visible range of the spectrum, while using different sized milling beads for dispersing. At a given light wavelength, the transmission improves with decreasing milling bead size, starting from 2 mm diameter

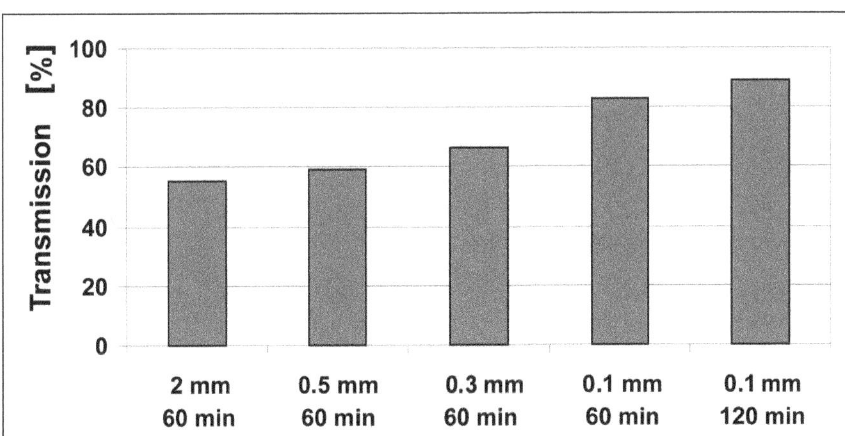

Figure 5.17: Transmission of nano-scaled TiO_2-pastes after dispersion with milling beads of different sizes. The concentration of TiO_2 was 31.2 per cent by weight in all cases.

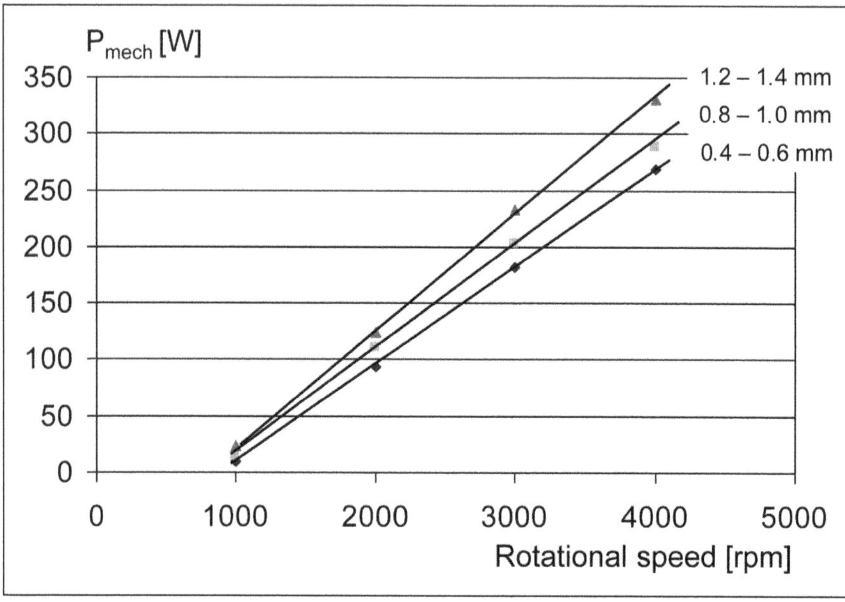

Figure 5.18: Dependency of the mechanical power on the rotational speed of the shaft for milling beads of different sizes

(mean value for a size range between 1.6 and 2.4 mm), over 0.5 mm, 0.3 mm down to 0.1 mm bead size. By doubling the dispersion time from 60 minutes to 120 minutes while using 0.1 mm, the transparency is further enhanced.

According to the dispersion Equation 5.11, improved dispersion results are expected if either the stress probability, breaking probability or both are increased. Should the milling beads have an influence on the breaking probability, then one would expect larger beads to be more efficient. It is well known that hard agglomerates require a certain minimum bead size since the kinetic energy of milling beads is directly proportional to their mass. Figure 5.18 shows the mechanical power of a bead mill (Dispermat SL, VMA Getzmann, Reichshof, Germany) as a function of the rotational speed for different bead sizes employed in such way that the free volume in the milling chamber was identical in all cases.

Figure 5.18 shows that smaller milling beads lead to lower mechanical powers at a given rotational speed than larger milling beads do. Thus, the breaking probability (see dispersing Equation 5.11) is expected to be even smaller. Yet, if dispersing results are improved using smaller beads, then the stress probability must have been augmented. How can this be understood?

In the case of nano-scaled pigments, not only the primary particles are very small, but the agglomerates that are to be dispersed, also. This becomes more difficult since, on the one hand, at a given shear rate, the difference in the velocity of the liquid phase on two adjacent sides of an agglomerate becomes lower, the smaller the agglomerates get. This lowers the effective volume per time unit. On the other hand, smaller agglomerates not only have a smaller mass, but also less inertia in comparison to larger agglomerates. For that reason they are more easily swept away out of the effective volume. Two milling beads on a collision course displace the liquid in between them before they are actually able to stress the agglomerates. The agglomerates may then just flow away together with the mill base. A denser packing of milling beads lets the agglomerates encounter more stress situations that they cannot avoid so easily. Exactly that is the reason why the efficiency of dispersions increases with the amount of milling beads also (see Figures 4.11 and 5.16).

5.4 Mechanical power and dispersion results in bead mills

The mechanical power is essentially the product of the rotational speed n between a shaft and its torque M, *as can be seen from Equation 5.13.*

The same mechanical power may therefore be introduced into a milling chamber either at higher rotational speed and lower torque or vica versa (2 · 4 = 4 · 2). Whereas the rotational speed is a pure machine parameter, the torque depends upon the flow properties of the mill bases. Standard mill bases often show very temperature dependent viscosities. This is discussed further in the Chapters 5.3.4 and 3.5.1.

For high speed bead mills, the breaking probability is a distinct function of the mechanical power employed [8], irrespective of the relative contributions of rotational speed and torque to it. Figure 5.19 shows gloss development curves for a mill base of an anti-corrosive paint dispersed in a 1.4 litre pin disc mill (PM-STS 1, Drais, Mannheim) in dependency of the rotational speed. As described earlier, paint samples were taken when the quasi-stationary power equilibria were reached.

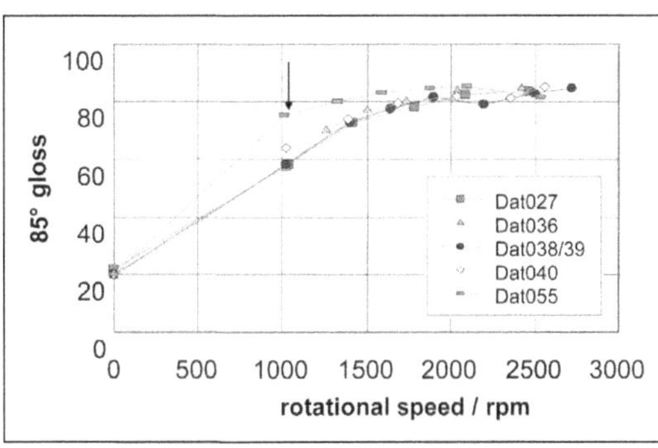

Figure 5.19: Gloss of a mill base in dependency of the rotational speed. DAT027 to DAT040: Experiments performed to check reproducibility. DAT055: Experiment in which the cooling water flow rate tripled.

An attempt was made to keep all the relevant parameters effecting the dispersions constant with the exception of the experiment denoted Dat055. In that dispersion the cooling water rate was tripled in comparison to the other dispersions. Figure 5.19 shows that the dispersion result improved comparatively in the low rotational speed range by this measure. A plot of the same gloss values as a function of the respective mechanical powers is presented in Figure 5.20.

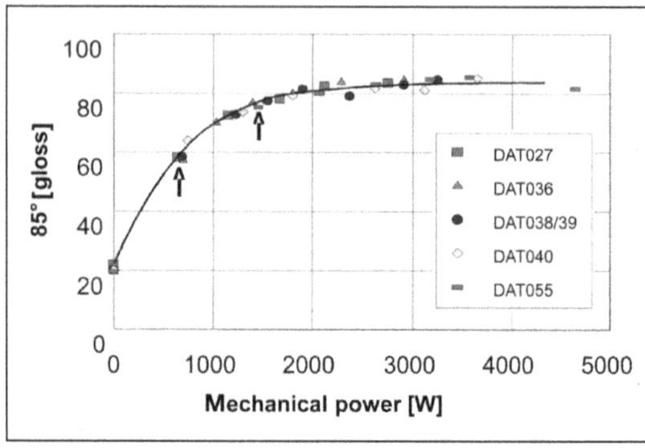

Figure 5.20: Gloss of a mill base in dependency of the mechanical power. DAT027 to DAT040 symbolize experiments performed to check reproducibility. At DAT055: the cooling water flow rate was tripled. The red arrows depict dispersions carried out at the same rotational speed of 1020 rpm.

In this plot, all gloss values fall into a single, combined curve. For the experiment with the extensive cooling, the mechanical power corresponding to the lowest possible rotational speed of 1020 rpm is 1400 W instead of 700 W because of the higher torque of the shaft. Compared to the other dispersions, due to the improved mechanical breaking of the agglomerates, the 85° gloss increases by almost 20 units. The same dependency was found for the colour strength development [9].

The distinct functionality between the absolute value of the mechanical power input and the achieved dispersion is very important in view of controlling dispersion processes. This functionality is probably due to the circumstance that, according to Equations 4.1 and 4.2, the deagglomeration is proportional to the product between the shear rate and the shear stress within a mill base. In first approach, the shear rate is proportional to the rotational speed of the shaft, whereas the shear stress is proportional to the torque on the shaft. At a low viscosity and high rotational speed, the agglomerates are loaded with a higher shear rate. At a low rotational speed and high viscosity, the shear stress becomes more important.

5.5 Literature

[1] J. Winkler, Farbe + Lack 89 (1983) 332–336
[2] J. Winkler, "Titanium Dioxide", Vincentz Network (2003) p. 50–53
[3] D. v. Pigenot, VII Fatipec-Congress, Vichy (1964) Congress brochure, p. 249
[4] M. J. Smith, J. Oil and Colour Chemists' Assoc. 56 (1973) 165
[5] O. J. Schmitz, R. Kroker, P. Pluhar, Farbe + Lack 79 (1973) 733
[6] A. Klaeren, H. G. Völz, Farbe + Lack 81 (1975) 709
[7] U. Zorll, Farbe + Lack 80 (1974) 17
[8] J. Winkler, E. Klinke, L. Dulog, J. Coatings Technology, 59, No. 754 (1987) 35–70
[9] J. Winkler, PhD Thesis, University Stuttgart, 1983

6 Stabilization against flocculation

Next to the wetting of the pigments and extenders and their mechanical breakage, the stabilization of the achieved particle distribution against flocculation is an essential step in the dispersion process.

Given that the mill bases or the finished, completed paint formulations respectively, have no yield points, the pigment particles will move around in the suspensions due to the Brownian motion of the liquid molecules. Even if the particles had no attractive forces amongst each other, they would randomly collide sooner or later. But, since they are attracted to one another (see Chapter 2.3), every encounter between the particles would let them stay together again. This re-agglomeration of dispersed particles is called "flocculation".

If flocculation occurs or not depends widely upon the physicochemical conditions at the interface between the particles and the liquid phase. These interactions, namely the flocculation or the stabilization of colloidal suspensions, are the main concern of colloidal chemistry. This field of science investigates the nature and the distance dependency of attractive and repulsive forces between the particles. At any particle-particle separation distance, the sum of attraction and repulsion determine whether the net force on the particles is attractive or repulsive, i.e. whether they flocculate or remain stable. Chapter 6 deals with these interactions.

6.1 Flocculation kinetics

6.1.1 Spontaneous flocculation kinetics

The model of *Smoluchowski* [1] is normally used to estimate the rate of flocculation of colloidal particles when no stabilization is in effect at all. Smoluchowski proposed a second order kinetic for describing diffusion controlled ("perikinetic") flocculation. The mathematical approach is that the decrease in the number of particles that haven't suffered a collision is proportional to the time t and to the square of the number of particles N present at any instant. The proportionality constant is k_F.

Equation 6.1 $$-\frac{dN}{dt} = k_F \cdot N^2$$

Jochen Winkler: Dispersing Pigments and Fillers

ISBN 978-3-86630-852-7

Equation 6.1 represents the collision frequency of the colloidal particles under the following simplified assumptions or conditions, respectively:

1) All primary particles are spherical in shape and of the same size
2) Every collision leads to the generation of an agglomerate
3) No more than two particles are involved in every collision
4) The agglomerates that are formed maintain their integrity

Integration of Equation 6.1 within the limits $t = 0$, where N_0 particles are present, to the time t, where N_t particles are not yet flocculated, yields

Equation 6.2
$$\frac{1}{N_t} - \frac{1}{N_0} = k_F \cdot t$$

According to Smoluchowski, the constant k_F is proportional both to R, the sum of the radii of the particles involved, and to the diffusion constant D.

Equation 6.3
$$k_F = 4\pi \cdot R \cdot D$$

The diffusion constant for round particles is expressed by

Equation 6.4
$$D = \frac{kT}{6\pi \cdot \eta \cdot r}$$

with

k = Boltzmann constant
T = absolute temperature
η = dynamic viscosity
r = particle radius

The Boltzmann constant connects the kinetic energy of molecules with the absolute temperature. It depicts by how many joules the energy of a molecule is increased when the temperature is raised by one K[1]. The Boltzmann constant is therefore related to the Brownian motion, which is the root cause for the diffusion of colloidal particles in suspensions. The expression in the denominator of Equation 6.4 describes the friction of a particles moving in a liquid. Thus, the diffusion is ultimately determined by the ratio of a propelling term (in the numerator of Equation 6.4) and a friction term (in the denominator of Equation 6.4) By inserting Equation 6.4 into Equation 6.3 and Equation 6.3 in 6.2 while assuming a the particles to be monodisperse (=of equal size), so $R = 2r$, yields

Equation 6.5
$$\frac{1}{N_t} - \frac{1}{N_0} = \frac{4}{3} \cdot \frac{kT}{\eta} \cdot t$$

[1] It is $1.38 \cdot 10^{-23}$ J/K

At the half time $t_{0,5}$, that is when half of the particles have flocculated, $N_t = N_0/2$. Inserting of $N_t = N_0/2$ into Equation 6.5 then leads to an expression from which the half time can be estimated.

Equation 6.6
$$t_{0,5} = \frac{3\eta}{4 \cdot kT \cdot N_0}$$

When no stabilization is in affect, flocculation occurs very quickly. In the case of a 10 weight% aqueous suspension of a titanium dioxide pigment, for example, a half time for flocculation of about 0.1 seconds is calculated from Equation 6.6 (pigment density 4.2 g/cm³, particle diameter 0.3 µm). However, Equation 6.6 is only a "snap-shot" of the flocculation process, since the flocculates consisting of two particles each continue to grow by colliding both with other agglomerates as well as with further primary particles. In the course of ongoing flocculation, the collision radius R increases, whereas the number of particles N decreases. Investigations showed, however, that the Smoluchowski theory is able to describe the behaviour of hydrophobic colloids nicely [2] in spite of all its simplifications.

Perikinetic flocculation takes place until the resulting agglomerates are in a size range of approximately 5 µm. A flocculate of titanium dioxide pigment particles of such a size is made up of more than 1500 primary particles (estimated by assuming the flocculates to be spherical and only a third of their volume consisting of pigment particles). Flocculates of that size no longer diffuse, but settle instead[2]. In doing so, further collisions occur since larger particles settle faster than smaller ones. Stirring of suspensions can cause the formation of even larger flocculates that tend to be less stable against shear so that they are more easily re-dispersed. Flocculation caused by stirring of suspensions plays a great role in waste water treatment. This type of flocculation is called "orthokinetic". The interplay of the various mechanisms, namely perikinetic flocculation, orthokinetic flocculation, flocculation by sedimentation and perhaps flocculation and/or re-dispersion caused by stirring is all in all very complicated, so that a mathematical modelling is far from being easy.

It can be concluded that pigment suspensions that are not stabilized in any way flocculate very quickly. As a rule, when this happens in polymeric composite materials, the quality is affected in a very negative way. Thus, the colour strength, the gloss as well as the gloss retention of paints films upon weathering are reduced. Furthermore, mechanical paint properties are not as good when the pigment particles are flocculated.

[2] Sedimentation: refer to Chapter 6.3

6.1.2 Measurement of flocculation rates

There are a number of possibilities for measuring the kinetics of flocculation. One way is to determine the increase in the weight of a sediment for undelayed m_{max} and delayed m_t flocculation with a sedimentation balance and plot the relative weight m_t/m_{max} against the time. This method yields S-shaped curves. As a measure for the mean flocculation rate the time can be determined, at which a certain percentage, say for example 40 % (($m_t/m_{max} = 0.4$) has settled [3]. A further possibility can be to consider the flocculation rate to be related to the speed with which a pigment free zone is formed on the surface of a suspension [4]. Especially diluted systems allow for measuring the transparency as a function of the time. The tangent of a plot of transparency versus time at low time lengths may be taken as a measure for the flocculation rate [5].

Using particle size analyzers whose measuring principles are based on laser light scattering, it is possible to determine particle size distributions very quickly. In that case, the change in the mean particle size may be determined as a function of time and taken as a measure for the flocculation rate [6].

6.1.3 Delayed flocculation

Provided that the pigment particles do not flocculate instantaneously, but that some mechanisms lead to their stabilization, this will be established by measuring their flocculation rate. The flocculation rate of the particles that are stabilized to some extent divided by the unperturbed flocculation rate is a relative measure for the efficiency of the collisions to form flocculates. This collision efficiency factor α represents the fraction of collisions that produced larger entities. In the case of delayed flocculation, Equation 6.1 can be modified to

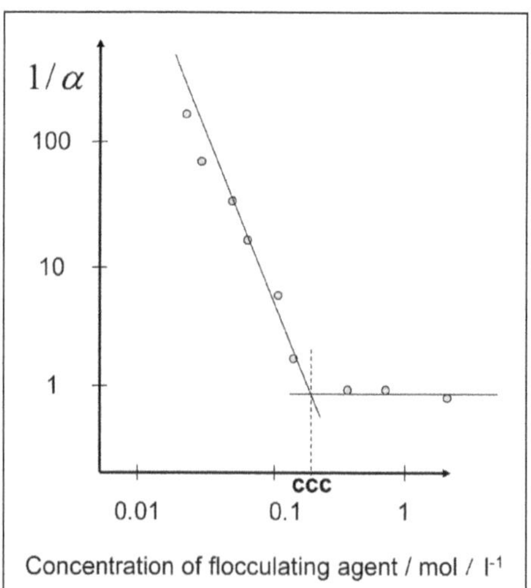

Figure 6.1: Determination of the critical coagulation concentration

Equation 6.7 $$-\frac{dN}{dt} = \alpha \cdot k_F \cdot N^2$$

The collision efficiency factor then stands in the numerator of modified Equations 6.2 and 6.5, whereas it is in the denominator of a modified Equation 6.6 and can obtain values ranging between zero and one. Following *Fuchs* [7], the reciprocal value of α is called "stability ratio". The relationship between the stability ratio and the stabilizing energy is discussed in Chapter 6.3.

On many occasions, it is of interest how the flocculation rate changes, for example, upon the addition of a flocculation reagent and at which concentration the change from a delayed to an undelayed, spontaneous flocculation occurs. For that, the logarithm of the stability ratio is plotted against the logarithm of the additive concentration. Two sets of data points are generated that belong to two straight lines. Their intersection defines the critical coagulation concentration "ccc" (see Figure 6.1). The ccc denotes the concentration of flocculating agent at the onset of rapid flocculation.

6.2 Sedimentation

Provided that the pigment particles are mobile in the liquid phase, that is to say that there is no yield point in the suspension, they will inevitably sink to the bottom sooner or later. The only exception are particles of only a few nanometres in size that, depending upon their specific gravity, may be so light that they are held in suspension by the Brownian motion of the liquid molecules.

The rate at which round particles settle is described by the Stokes equation. The particles are subjected to the gravitational force F_G on one side and the buoyancy force F_B, that is caused by the difference in specific gravity of the particle and the medium, on the other side. If the particles have a greater density than the liquid, then they settle as a result of the net force. Yet, in doing so, they are slowed down by the frictional force F_F. Particles of a diameter d settle with a constant speed v as soon as the frictional force is in balance with the gravitational force reduced by the buoyancy.

Equation 6.8 $$F_F = F_G - F_B$$

According to Stokes, the frictional force is given by

Equation 6.9 $$F_F = 3 \cdot \pi \cdot \eta \cdot d \; v$$

The gravitational force on a particle having the volume V_P and the density ρ_P is

Equation 6.10 $$F_G = \rho_P \cdot V_P \cdot g$$

with the gravitational constant g (g = 9.806 m/s²). The buoyancy of a particle in a liquid of density ρ_F is

Equation 6.11 $$F_B = \rho_F \cdot V_P \cdot g$$

Inserting Equations 6.9 to 6.11 into Equation 6.8 and rearranging leads to an expression for the settling rate.

Equation 6.12 $$v = \frac{V_P \cdot g\,(\rho_P - \rho_F)}{3 \cdot \pi \cdot \eta \cdot d}$$

Assuming the particles to be spherical ($V_P = 1/6 \cdot \pi \cdot d^3$), Equation 6.12 turns into Equation 6.13.

Equation 6.13 $$v = \frac{d^2 \cdot g}{18 \cdot \eta} \cdot (\rho_P - \rho_F)$$

Equation 6.13 is valid only for slow sedimentation rates for which the Reynolds numbers[3] are smaller than unity. Furthermore, the viscosities are not allowed to be too high if the settling velocity is to be described by the Stokes equation. Larger particles settle slower than calculated from Equation 6.13 [8–10]. Figure 6.2 shows sedimentation velocities of sand particles of density 2.65 g/cm³ in water, calculated from a formula of *Soulsby* [8], in a double logarithmic plot. Likewise, the sedimentation rates calculated from the Stokes Equation 6.13 are plotted. There is a good agreement between the Stokes and the refined Soulsby equation up to particle diameters of approximately 0.1 mm (100 µm).

Therefore, the Stokes equation is good enough for describing the settling behaviour of pigments, especially since 100 µm large particles settle very quickly, as can be seen from Figure 6.2.

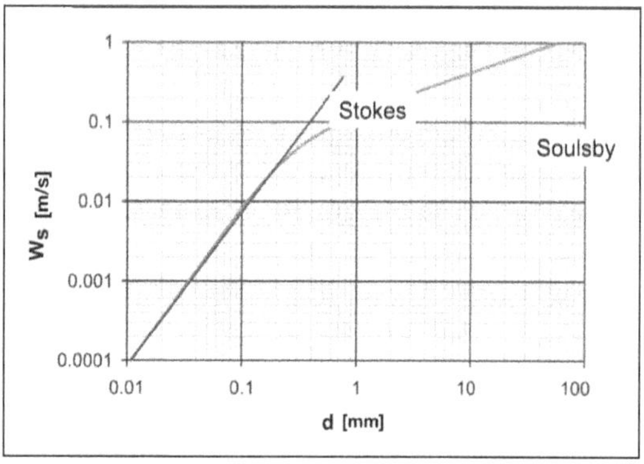

Figure 6.2: Settling rate W_s of sand kernels of diameter d and density 2.65 g/cm³ in water at 20 °C, calculated from a formula of Soulsby [8]. Original diagram taken from Wikipedia. Search word: "Sedimentationsgeschwindigkeit", engl. sedimentation rate. The graph was complimented with the sedimentation rate from the Stokes equation by the author.

The degree of flocculation has a large effect on the height and the density of the sediment formed. When settling occurs from a status of good stabilization, that is to say when the particles settle individually, then the settling takes a long time, yet a thin and confirmed sediment is formed that is difficult to re-disperse. If, however, flocculation occurs, then the pigments settle quickly to yield bulky sediments that are easily stirred up again. The reason is that, when flocculation occurs, the pigments form

[3] The Reynolds number is a dimensionless number that represents the quotient between inertial forces and viscous forces in fluid dynamics.

loosely packed flocculates which inherit a great deal of liquid in their structure. The more bulky the flocculates are, the easier it is to re-disperse the pigments by mere stirring. The sediments may be characterized by measuring the work that is necessary to press a test body, such as a rod, into them. For that, the force of the entering rod is measured as a function of penetration depth and the integral over the whole height of the sediment is formed.

From applications point of view, both extremes, slow flocculation to a compact sediment as well as very rapid flocculation to a loosely packed sediment, are undesirable. Ideally, the systems should be able to be re-dispersed by stirring, yet should be stable against flocculation for a longer period of time than needed for their processing. Additives are available on the market that lead to controlled settling due to "bridging flocculation".

6.3 Potential curves

As discussed in Chapters 2.4 and 2.5, pigment particles always attract one another by the London-van der Waals interaction. The attractive force is distance dependent and, in case of pigment particles, is relevant for separation distances close to zero up to distance ranges comparable to the dimensions of the pigment particles. When two pigment particles approach each other to such a distance by chance, then flocculation, or, when in air, agglomeration takes place. In order for pigments to be stabilizes against flocculation, some other opposing forces have to come into affect that have similar reaches. The pigment particle distribution will then depend upon the balance of attractive and repulsive influences at every separation distance. Normally, attractive and repulsive energies are compared. For historical reasons, it has become customary to relate to "potential energy curves" (or merely "potential curves"). The surface separation distance dependent total potential curve V_{tot} is the sum of the attractive energy V_{attr} and the repulsive energy V_{rep} at any separation distance.

Equation 6.14

$$V_{tot} = V_{attr} + V_{rep}$$

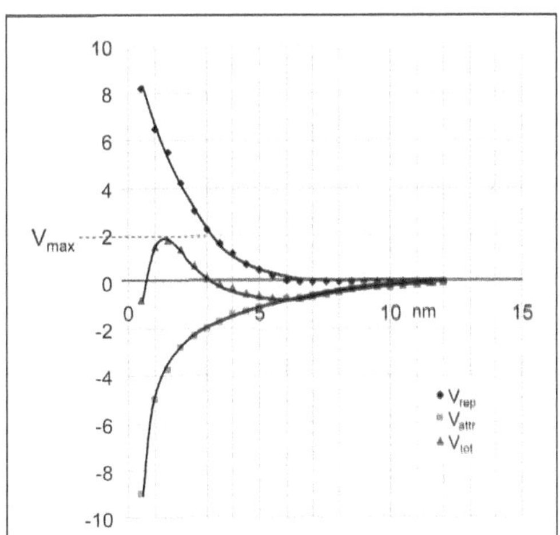

Figure 6.3: Potential energy curve V_{tot} as the sum of an attractive energy curve V_{attr} and a repulsive energy curve V_{rep} (schematically)

Figure 6.3 shows the principle. The surface separation distance of the particles is plotted on the x-axis, whereas the attractive energy, the repulsive energy and the resulting total energy are plotted on the y-axis. The energy of attraction gets a negative sign and the energy of repulsion a positive one. This example shows the energy of attraction reaches farther from the particle surface than the repulsive energy. For that reason, attraction prevails at larger separation distances. If the repulsive energy grows faster than the attractive energy as the particles approach each other, then repulsion prevails and the potential energy curve becomes positive. At still further approach, the attraction may once again become dominant and therefore the potential energy curve goes through a maximum before declining sharply. The part of the potential energy curve between a noticeable onset of attraction at large separation distances and the point where V_{tot} becomes positive for the first time is called the "secondary minimum". It is a shallow potential basin in which the particles flocculate loosely. The particles need only little kinetic energy to separate from one another again so that the particles must be sheared only lightly, for example by mere stirring, in order to be re-dispersed.

Depending upon the height of the energy barrier V_{max} which follows upon further approach, the particles are more or less well kept from flocculating with each other. If they have enough kinetic energy to override the potential energy barrier, then they may fall into the steep "primary minimum" where attraction is very predominant. In that case, simply stirring will not suffice to re-disperse the particles again. That's why storage stability testing is normally done at elevated temperatures. The fraction of the pigment particles with a higher kinetic energy (Maxwell-Boltzmann distribution) becomes larger when the temperature is high so that it is possible to predict in a shorter time if a paint formulation will remain stable even when stored for a long time. Since the kinetic energy of particles is a distributed quantity, there is no sharp distinction between stable and unstable systems. There are only formulations in existence in which the pigment particles flocculate more or less quickly. It should therefore be mentioned that the potential energy curves can have just about any shape imaginable. When a repulsive action is low or missing completely, the total potential curve may be attractive only. On the other hand, there might be no primary potential minimum at all or it may be very shallow.

According to *Fuchs* [7], the total potential energy curve may be related to the collision efficiency factor α of delayed flocculation (Equation 6.7). The stability ratio W, that was already mentioned, is given by the ratio of the rate constants of instantaneous and delayed flocculation (see Equations 6.1 and 6.7).

Equation 6.15 $$W = \frac{k_F}{\alpha \cdot k_F} = \frac{1}{\alpha}$$

Since α becomes smaller the more stable the pigments are in a system, W obtains large values when flocculation is inhibited. According to Fuchs [7], the stability ratio may be expressed as a function of the total potential energy curve

Equation 6.16 $$W = \frac{1}{\alpha} = 2r \int_{R=2r}^{\infty} \frac{1}{R^2} \cdot \exp\left(\frac{V_{tot}}{kT}\right) \cdot dR$$

whereby r is the radius of the (spherical) particles and R is their centre to centre distance.

In Equation 6.16, the exponent of the total energy is expressed as a multiple of kT. For a colloidal suspension to be reasonably stable, the energy barrier V_{max} of the total potential energy curve should be ten to fifteen times kT [11, 12]. Following *Verwey* and *Overbeek* [11], an energy barrier of fifteen times kT slows the rate of flocculation by a factor of 10^{-5}. Other authors [13] consider an energy barrier between 5 kT and 10 kT to be sufficient.

The reason for this uncertainty lies in the fact that the total energy curve V_{tot} and thus V_{max} from Figure 6.3 cannot be determined directly, just as V_{attr} and V_{rep} can't. It is merely possible to estimate V_{attr} and V_{rep} using the theory of the attraction between particles from *Hamaker* (Chapters 2.4 and 2.5), the DLVO theory for the stabilization of particles by electrostatic charges (Chapter 6.4.1) or the theories for the stabilization of particles by adsorbed polymers (Chapters 6.5 ff.) and to calculate V_{tot} from that. The virtue of these theories is not to predict the stability of colloidal systems but rather to obtain a principle understanding of the factors determining the stability against flocculation.

6.4 Electrostatic stabilization

The fundamentals of stabilizing colloidal particles against flocculation by electrostatic forces were studied independently from one another by *Derjaguin* and *Landau* [14, 15] from Russia and by *Verwey* and *Overbeek* [11] from the Netherlands in the forties of the last century. In their honour, the resulting theory is called the DLVO theory. The DLVO theory describes the interactions between charged colloidal particles in suspensions and relates these with the van der Waals attraction.

6.4.1 Electrostatic charging of pigment particles

The electrostatic stabilization of colloidal particles relies upon particles in a suspension having ionic charges. If these charges are of the same sign, then the particles repel each other, whereas oppositely charged particles attract each other. In principle, three different mechanisms can lead to these charges.

a) adsorption of ions from the solution
b) acid-base interactions with the liquid components in a suspension
c) adsorption of polyelectrolytes

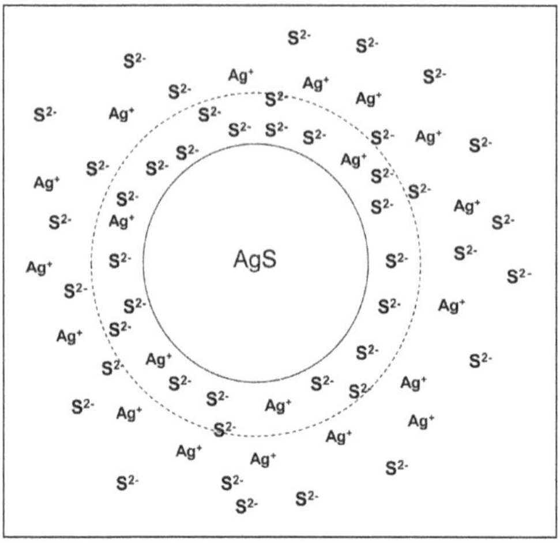

Figure 6.4: Charging of a silver sulfide crystal by the favoured adsorption of sulfide ions

d) adsorption and interaction with other electrostatically charged particles in a suspension
e) special cases in organic systems

a) Adsorption of ions from the solution

Ions come into being when molecules dissociate. When that happens, just as many positively as negatively charged ions are generated. This is referred to as the "electroneutrality principle" in a solution. Positively charged ions are called cations and negatively charged ions are named anions. If one of the ion species is less soluble than the other, or if the pigment surface has a special affinity to one of them, then this ion type will accumulate on the pigment surface and become "potential determining". The counter ion is then still within the suspension, yet not spatially concentrated in any way. Figure 6.4 is a schematic representation of this situation.

Also, organic ions such as the anions of surfactants can adsorb onto pigment surfaces and become potential determining. Examples for this are the salts (soaps) of alkyl benzene sulfonic acids or the soaps of fatty acids in aqueous solutions. In both cases, an adsorption of the anions via their head groups, which is the hydrophilic, ionic group, seems to occur initially. Upon further adsorption, the ionic groups of the adsorbates[4] reach into the solution. In the case of the adsorption of alkyl benzene sulfonates onto titanium dioxide particles from aqueous solutions, different parts of the adsorption isotherms may be distinguished in which the molecules point into different directions [16]. When fatty acid anions adsorb onto barium sulfate particle surfaces, for example, the particles first become water repellent since the carboxyl groups orientate towards the surface whereas the alkyl chains are located outwards, away from the particle surfaces. When the surface is completely covered with a monomolecular layer, further adsorption occurs in such a way that the non-polar parts of the first adsorption layer interact with the non-polar parts of the second layer so that the ionic groups are now positioned to the outside and the particles become hydrophilic [17]. Fig-

[4] A moiety that attaches onto an adsorbent is called an adsorbate.

ure 6.5 depicts that schematically. Since, normally, that part of an amphiphilic molecule orientates towards the surface, that is less soluble in the liquid phase, the primary step in this case will be chemisorption[5] of the adsorbates.

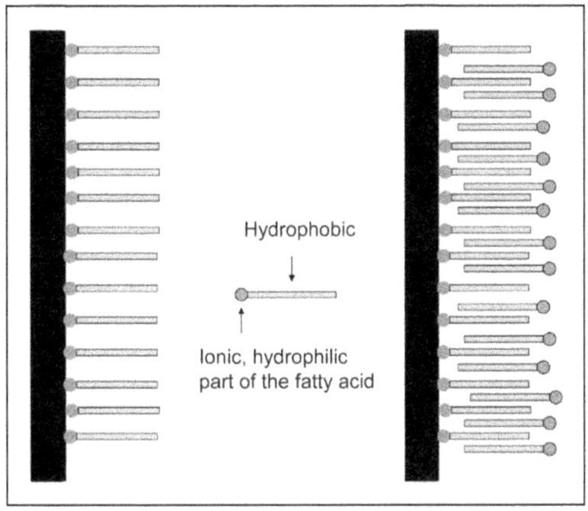

Figure 6.5: Electrostatic charging of pigment surfaces by the adsorption of surfactants. Left: monomolecular layer renders hydrophobic particles. Right: bimolecular layer renders hydrophilic particles

b) Acid-base interactions with liquid components

Acid-Base interactions are limited to amphoteric pigments. Amphoteric pigments can react with acidic reaction partners as bases and with basic components as acids. All pigments consisting of metal oxides, sulfides or nitrides are amphoteric in nature since they have -OH, -SH or =NH groups, respectively, on their surfaces. These groups are present because the electroneutrality principle is in effect in ionic crystals as well. Otherwise, the crystals would be electrostatically charged. In an oxidic pigment crystal, for example, the negative charges stem from O^{2-} ions whereas the positive charges come from Me^{n+} ions (Me = metal). Since oxygen ions always coordinate with at least two metal ions, oxygen anions and metal cations alternate in the crystals. If the crystals should terminate with a $Me\text{-}O^-$ group, then the negative charge is compensated by a proton H^+. If there is a Me^+ or Me^{n+} at the terminating surface, then the charge is compensated by one or more OH^- ions. In both cases Me-OH groups are formed. If pigments are synthesized from aqueous solutions, then protons and hydroxyl ions are abundantly available. But also when pigments are produced for example in gas phase synthesis, the moisture in the air is sufficient for these reactions to take place.

The surface concentration of the -OH groups depend upon the crystal structure of the pigment. Titanium dioxide in the rutile modification has about six -OH groups per nm^2, whereas for anatase, seven to eight -OH groups are reported [18]. Pyrogenic silicas, for comparison, have approximately two -OH groups per nm^2 [19], whereas precipitated aluminium oxides have somewhere between eight and nine -OH groups per nm^2 [20].

[5] Chemisorption or „activated adsorption“ is an adsorption in which chemical bonds are formed.

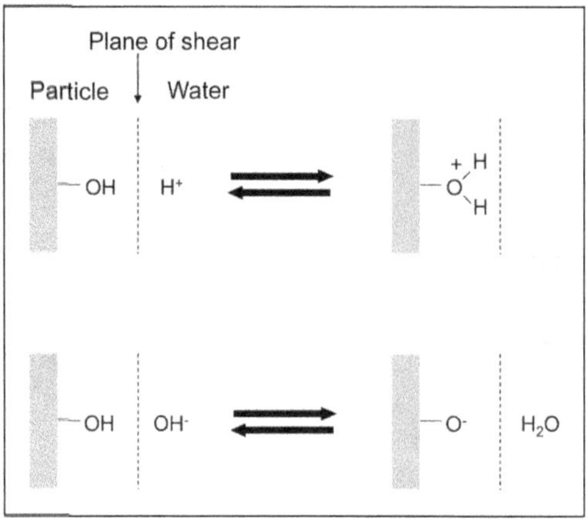

Figure 6.6: Reactions of hydroxyl groups on the surfaces of amphoteric oxides with acids and bases

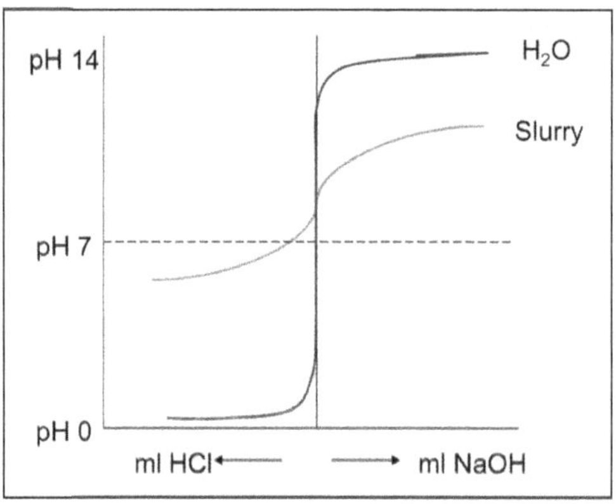

Figure 6.7: pH titration curve of pure water in comparison to the pH titration curve of an aqueous suspension of amphoteric pigments

The -OH, -SH and =NH groups are able to react with acids as well as with bases. Figure 6.6 shows the principle for the example of a metal oxide pigment. When, for example hydrochloric acid, is given to an aqueous suspension of the pigment (top half of Figure 6.6), an equilibrium is formed between protons that remain in solution and protons that connect to surface hydroxyl groups by hydrogen bridging. The adsorbed protons impart a positive electrostatic charge to the pigments. Otherwise, when a base, for example caustic soda (NaOH, see lower half of Figure 6.6), is added, the surface hydroxyl groups can lose a proton and react with the sodium hydroxide ions to form water. In that case, the pigments acquire a negative charge.

Apart from the positive and negative charging of the particles, Figure 6.6 also shows that the pigments themselves act as solid acids and bases. Basically, the reactions from Figure 6.6 are neutralization reactions. So, amphoteric pigments have pH-buffering capabilities. This is shown schematically in Figure 6.7.

The pH titration curve denoted "H_2O" is that of pure water. Starting ideally from a pH value[6] of 7, the pH increases strongly when a base is added. If acid is given to water, the pH declines just as sharply. The curve denoted as "particle suspension" in Figure 6.7 shows the influence of amphoteric pigment particles in a principle manner. When a base is added to the suspension of particles, the pH rises less and

[6] The pH value is the negative decimal logarithm of the hydrogen ion activity (concentration) and is connected with the dissociation constant of water. At a pH value of 7, 10^{-7} protons are in one liter of a liquid.

when acid is added, the pH value does not fall so sharply. pH titrations of particle suspensions provide a route to measuring the adsorption of protons and hydroxyl ions on amphoteric particles [21].

c) Adsorption of polyelectrolytes

Macromolecules with functional groups that may act as acids or bases are called "polyelectrolytes". Polyacrylic acid is an example for a polyelectrolyte. It has -COOH groups that may react with bases to form salts. When ammonia is added to a mixture of polyacrylic acid and water, the carboxyl groups saponify to $-COO^-$ and NH_4^+, which are soluble in water. Since the carboxyl group is part of the polymer molecule, the polymer itself becomes negatively charged. (In fact, the polymer only becomes water soluble by that reaction.) When such a polymer adsorbs to pigment surfaces, the pigments themselves become negatively charged.

Next to carboxyl groups, sulfonic acid groups, sulfate groups, phosphonate groups and phosphate groups can have the same effect when reacted into polymers. Inorganic salts such as sodium hexa metaphosphate ("Calgon") or potassium tripolyphosphate ("KTTP") are polyelectrolytes and charge colloidal particles negatively. Positively charged polyelectrolytes are seldomly used for pigment stabilization. Quite on the contrary, polyacrylic acid amides are widely used as flocculants in water treatment since they cause bridging flocculation [7] with the negatively charged, suspended matter. Examples of cationic polyelectrolytes with stabilizing effects are polymers with amine or amide groups evenly distributed along the polymer chain. Cationic electrodeposition coats, that are used for example in the automobile and the automobile supply industry, contain amine groups which are neutralized by organic acids such as formic acid.

d) Interaction with other electrostatically charged particles

Under certain circumstances, nano-scaled pigments and fillers can be electrostatically charged in solvent based formulations. In that case, they can assemble with other, coarser pigments, thus embossing their charge onto them. In that way, the coarser pigments are stabilized against flocculation [22].

Figure 6.8 shows a light microscopic picture of a paste of a nano-scaled barium sulfate in an alkyd melamine system (right side) that was brought into contact with a drop of flocculated Color Index[7] Pigment Violet 19 (quinacridone pigment, left side) in the same system. Without any mechanical agitation, a stabilization of the pigment particles comes into effect at the interface, which is demon-

[7] The Color Index system is a reference book of the „British Society of Dyers and Colourists" and the „American Association of Textile Chemists and Colorists", in which pigments are classified according to their chemical composition.

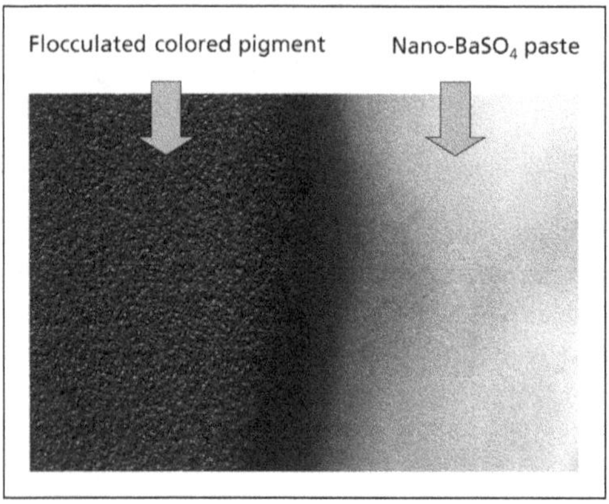

Figure 6.8: Photograph of the boundary between a drop of a flocculated suspension of Color Index Pigment Violett 19 with a paste of nano-scaled barium sulfate in a resin solution with polar solvents under a conventional light microscope

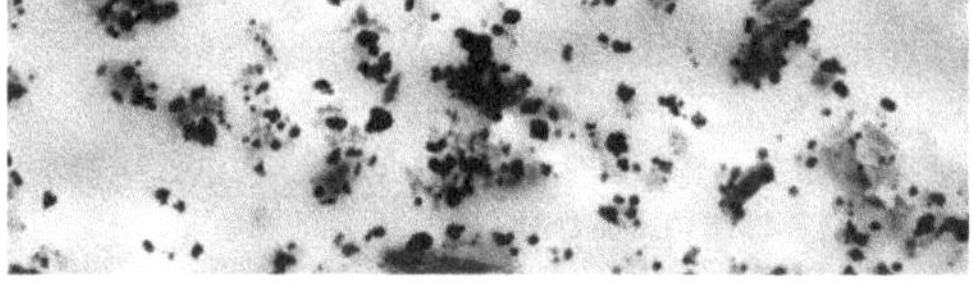

Figure 6.9: Distribution of the pigment particles in the cured polyester melamine formulation. Light grey: C. I. Pigment Violet 19; Dark grey: nano-scaled barium sulfate particles.

strated by the increase in colour strength. Figure 6.9 shows the pigment particle distribution in the cured paint film made from these components.

It is notable that each coloured pigment particle (light grey) is surrounded by a number of barium sulfate particles (smaller and darker in colour). Both nano-scaled aluminium oxides as well as nano-scaled titanium dioxides showed the same behaviour. The nano-particles had a positive charge in this system. It was shown for the barium sulfate nano-particles that the charges came from adhering sodium sulfate salt originating from the production process. The electrolyte dissociates partly in the system ($Na_2SO_4 \leftrightarrow Na^+ + NaSO_4^-$) and whereas the sodium sulfate anion is at least partly soluble due to its lower charge density[8] in comparison to the sodium cation, the sodium cation (Na^+) remains on the surface and becomes potential determining. Nanoparticles that were rinsed with water until free of electrolytes had no stabilizing effect. Conversely, it was possible to charge the nano-particles positively by adding an ethanolic potassium sulfate solution to the suspension. Nano-scaled barium sulfates are successfully used in some colour paste systems as flocculation stabilizers.

A similar stabilizing mechanism was found in the case of the adsorption of negatively charged polystyrene latex particles onto positively charged titanium dioxide pigments [23]. Depending upon the amount of polystyrene latex particles added,

[8] The charge density is the quotient between the number of charges and the volume of an ion.

the titanium dioxide which was originally stable due to its own charge, first flocculated, but, upon further latex addition, became stable again. This observation was interpreted in that the latex initially compensated the positive charge of the pigment particles, leading to flocculation. Upon further addition, entities were formed with a surplus of negative charges that were then stable once again.

e) Special cases in organic systems

The precondition for electrostatic charges to occur on pigment surfaces in organic media is the presence of a certain amount of polar solvents such as polyvalent alcohols (glycols), since the ions have to be dissolved and stabilized. A further way to induce charges in organic systems is to add fatty acid salts of divalent cations. The ions that are generated upon dissociation have a fairly low charge density, which enhances their solubility. The ion with the higher charge density then becomes potential determining because it is less soluble. Calcium octoate (Oct-Ca-Oct), for example, dissociates into an Oct^- anion and a $Ca\text{-}Oct^+$ cation. Since the octoate anion has a higher charge density than the calcium octoate cation, it is preferentially adsorbed. If the solubility of both types of ions is too pronounced, then, although dissociation into ions readily takes place, none of them go to the surface of the particles. The charging of particles in organic media is therefore characterized by the existence of delicate equilibria. The presence or absence of traces of moisture can already have a grave influence on the generation of charges [22].

Almost all pigments produced in an aqueous chemical procedure carry some sodium sulfate because in the course of their synthesis, pH-values have to be adjusted. For cost reasons, this is typically done with sodium hydroxide and sulfuric acid as reagents. When crown ethers are added to slurries of these pigments in organic solvents, the particles obtain a negative surface charge because the crown ethers form complexes with the sodium ions that migrate into the liquid phase whereas the anions remain on the surface [24]. Although this mechanism works in even the most unpolar media, due to the high price of crown ethers, it is unfortunately not economically viable for a technical use, but only of academic interest instead.

6.4.2 Fundamentals of electrostatics

Electrostatic charges are always a multiple of the charge of one electron. The physical unit for charges is coulomb *C*. One elementary charge has $1.6021765 \cdot 10^{-19}$ coulomb, or, 1 coulomb relates to $6.24 \cdot 10^{18}$ elementary charges, respectively.

Two oppositely charged elementary charges e_+ and e_- at a distance r in vacuum attract each other with the force F

Equation 6.17
$$F = \frac{e_+ \cdot e_-}{4\pi\varepsilon_0 r_2}$$

Equation 6.17 is known as Coulomb's law. In Equation 6.17, ε_0 is the "permittivity of vacuum", also called "electric field constant". $e_0 = 8.854 \cdot 10^{-12}\ C^2 \cdot N^{-1} \cdot m^2 = 8.854 \cdot 10^{-12}\ A \cdot s \cdot V^{-1} \cdot m^{-1}$. It is the dielectric constant of vacuum (see Chapter 1.2.1). Therefore, F has the physical dimension newton N.

Another important factor in connection with the charging of colloidal suspensions is the "electric field strength". The electric field strength E is defined as the force acting on a positive elementary charge in an electric field. From Equation 6.17 follows:

Equation 6.18
$$E = \frac{F}{e_+} = \frac{e_-}{4\pi\varepsilon_0 r^2}$$

E has the dimension newton per coulomb or volt per meter. $N \cdot C^{-1} = V \cdot m^{-1}$.

The still next expression that plays a role is the "electric potential". The electric potential Ψ is the work that is necessary to move an elementary charge from infinite distance, where the field strength is zero, into the vicinity of a charge of the same sign. Since E is the force on a charge in an electric field, the electric potential is given by

Equation 6.19
$$\Psi = \int_{\infty}^{r} E \cdot (-dr) = -\int_{\infty}^{r} E \cdot dr = -\int_{\infty}^{r} \frac{e}{4\pi\varepsilon_0 r^2} \cdot dr = \frac{e}{4\pi\varepsilon_0 r}$$

Equation 6.19 shows that the electric potential has the dimension volt.

6.4.3 Potential distribution surrounding an electrostatically charged particle

Figure 6.10 demonstrates the distribution of ions and the decay of the potential surrounding a particle with a positive surface charge in a schematic way.

The positive surface charge could, for example, stem from a protonation of surface hydroxyl groups or by the adsorption of cations from the solution. The positive charge is considered to be part of the particle itself and is depicted by a plus sign. The particle therefore has a surface potential, also called "Nernst potential", of Ψ_0. If a salt is dissolved in the liquid phase, cations and anions are generated. Due to the electrostatic interaction (Equation 6.12), the cations are repelled from the surface whereas the anions are attracted. This leads to the formation of a layer of negative charges directly on the particle surface. This first adsorption layer is called the Stern layer. Within the Stern layer, the potential drops linearly down to a value of Ψ_S since (almost) only one type of ion is present, namely anions in the case of a positive surface charge or cations, if the particles happen to be charged negatively. The concentration of ions in this layer results from the interplay between the degree of electrostatic attraction to the surface, the solvation[9] of the ions and the repulsive forces between the ions in

[9] The interaction between the ions and the solvent molecule is called solvation. When the solvent used is water, then this is called hydration.

the Stern layer. That is why not all of the surface potential is compensated by the Stern layer. Instead, a further layer follows in which the ions are bound less firmly so that they are able to diffuse within the liquid phase. In this "diffuse" part of the charge distribution, counterions at first prevail. Yet, at a farther distance from the particle surface, theoretically at an infinite distance from the surface, the number of cations and anions is equal so that the electrical potential is zero there. Both layers, the Stern layer and the diffuse layer together form the "electrostatic double layer". Figure 6.11 shows the electrostatic double layer surrounding a particle in a two dimensional graph. The particles are surrounded by a „cloud" of ions.

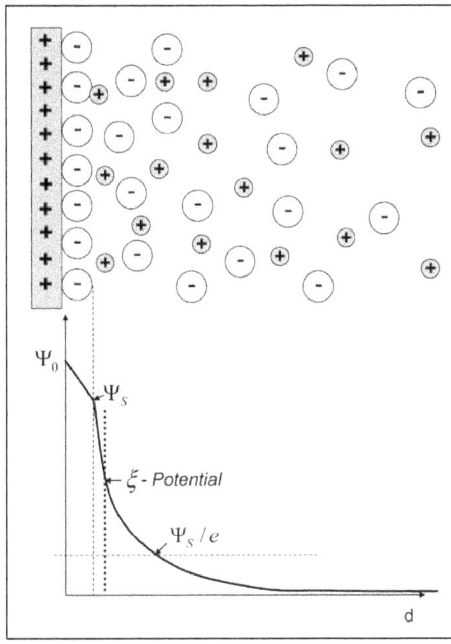

Figure 6.10: Potential distribution around a positively charged particle in an electrolyte solution (=solution of a salt in water)

According to *Guoy* [25], the decay of the potential in the diffuse part of the double layer can be approximated by a simple exponential equation, provided the potentials are not too high [26].

Equation 6.20 $\Psi_d = \Psi_s \cdot e^{-\kappa d}$

Ψ_d is the potential at a distance d from the particle surface. κ is called the "Debye-Hueckel parameter" or the "inverse Debye length" (see Chapter 6.4.4) and is expressed by

Equation 6.21

$$\kappa = \sqrt{\frac{4\pi \cdot e^2 \cdot \sum c_i \cdot z_i^2}{\varepsilon_r \varepsilon_0 \cdot kT}}$$

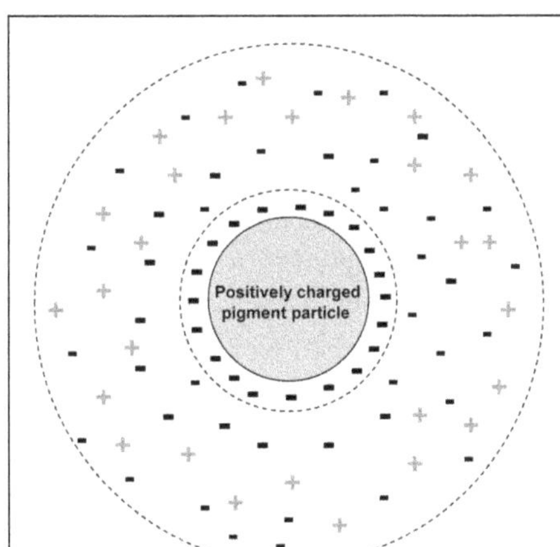

Figure 6.11: Ion cloud surrounding a positively charged particle

In Equation 6.21, ε_r is the relative dielectric constant of the liquid medium and $\sum c_i \cdot z_i^2$ the so called "ionic strength". The ionic strength is the sum

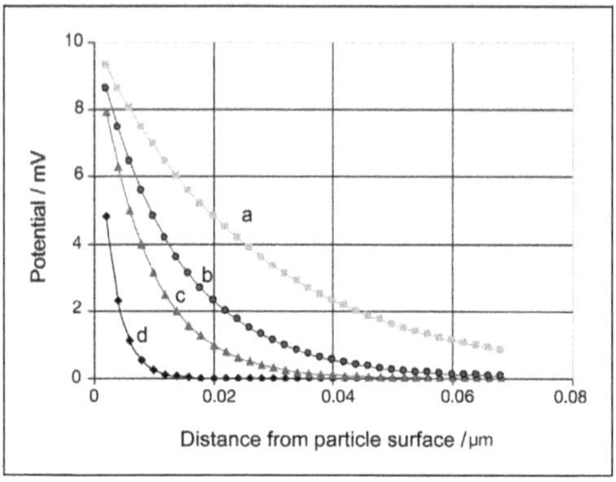

Figure 6.12: Potential surrounding a particle with a surface potential of 10 mV in a solution containing
a) 1:1 electrolyte, 10^{-5} mol/l
b) 2:2 electrolyte, 10^{-5} mol/l
c) 1:1 electrolyte, 10^{-4} mol/l
d) 1:1 electrolyte, 10^{-3} mol/l
Calculations from equation 6.20.

of the products of the concentrations *c* and the square of the ionic charges *z* of all types of ions present. All the other quantities in Equation 6.21 have the usual meanings.

According to Equations 6.20 and 6.21, the distribution of the potential depends greatly upon the concentrations and the valences of the ions in solution. Figure 6.12 shows the course of potentials in dependence of the distance from a particle surface for different concentrations of 1 : 1 and 2 : 2 electrolytes, respectively[10].

Figure 6.12 clearly shows that the potentials reach less far with increasing ionic strengths. As a consequence, when two charged particles approach each other in a medium of higher ionic strength, the onset of repulsive electrostatic forces is not until shorter surface separation distances where the van der Waals attraction may already be pronounced. This is the principle behind the flocculation of suspended matter from waste waters by the addition of high valence salts. Well known flocculating agents are copperas (iron-II-sulfate heptahydrate, which is gained as a by-product in the production of titanium dioxide pigments), Iron-III-chloride or poly aluminium chloride $((AlCl_3)_n)$, which is produced solely for that purpose. According to the Schulze-Hardy rule, different amounts of monovalent, divalent and trivalent ions are needed for rapid flocculation of colloids. In fact, the required amounts are inversely proportional to the sixth power of the valence, so that the ratio is about 1 : 0.015 : 0.0013. The ionic strength also plays a vital role in aqueous paint formulations in view of the stabilization of pigments and extenders against flocculation. That is why water based paints are produced with distilled or at least with cation exchanged water. The conductivity should be less than $20\,\mu S \cdot cm^{-1}$. Neither should the pigments and fillers themselves carry appreciable amounts of salt as a freight with them. Therefore, it has become a rule of the thumb that the conductivity of a 10 % pigment slurry in deionized water should have a conductivity of less than $100\,\mu S \cdot cm^{-1}$. The substitution of zinc as a dopant for titanium dioxide rutile pigments for improving weather

[10] A 1:1 electrolyte is a salt consisting of a cation and an anion which both have the valence one. In a 2:2 electrolyte, both anion an cation carry two electrostatic charges.

stability against aluminium was, for example, enhanced by the expansion of water based paints. The solubility of zinc from the pigments led to flocculation effects. Even if the ionic strength in an aqueous formulation can in general be higher without causing problems, it should be kept in mind that the concentration of the dissolved salts increase as the water evaporates from the applied paint films. In case the pigments are still mobile when this happens, the flocculation of the pigments can commence.

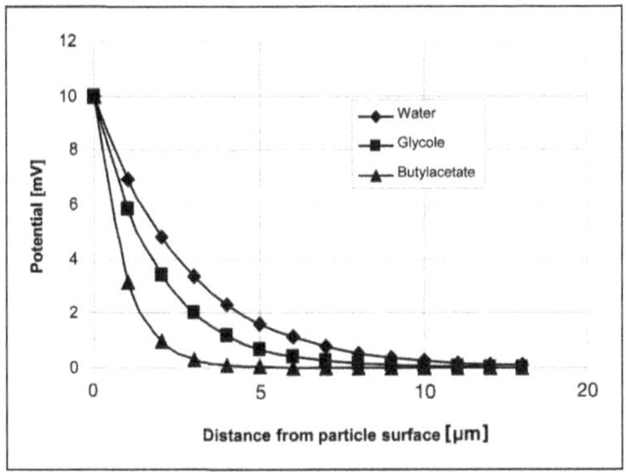

Figure 6.13: The potential distribution around a particle with a surface potential of 10 mV in a 10^{-3} molar solution of a 1 : 1 electrolyte
a) butyl acetate
b) glycole
c) water

According to Equations 6.20 and 6.21, the dielectric constant of the medium also has an influence on the potential distribution. Since ε_r is in the denominator of Debye-Hueckel parameter (Equation 6.21), the potential in organic media is expected to reach less far than in water, for comparison. This is shown in Figure 6.13 for water ($\varepsilon_r = 80$), glycole ($\varepsilon_r = 37$) and butyl acetate ($\varepsilon_r = 5$) at a concentration of 10^{-3} moles/litre of a 1 : 1 electrolyte. However, since salts are less soluble in media with low dielectric constants, there will be considerably less ions present. This is however true both for the ions that lead to the surface potential as well as for the ions in solution. If electrostatic charges are generated in organic media, then only few charges are sufficient to cause very far reaching potentials because of a lack of shielding ions in the diffuse layer.

6.4.4 Zeta potential

There is no way of measuring the surface potential of colloidal particles directly. Usually, the so called zeta potential[11] is determined by electrophoretic measurements instead. In electrophoresis, the migration speed of particles in an electric DC (direct current) field is measured. The "electrophoretic mobility" is defined as the migration speed divided by the field strength of the DC field. It is normally reported in the dimension $\mu m \cdot s^{-1}/V \cdot cm^{-1}$, whereby the speed of migration is normally measured in the order of only a few micrometer per second.

When electrostatically charged particles migrate in an electric field, they take a part of their electrostatic double layer with them. The ions that are closest to the particle

[11] Named after the Greek letter zeta ζ

surface have the strongest interaction with them. The farther the distance to the surface is, the less these interactions get. At some distance from the particle surface there is a "plane of shear" that denotes the distance at which the ions between this plane of shear and the particle surface migrate with the particle, whereas those outside of the shear plane stay behind. Therefore, the electrophoretic mobility that is measured in an experiment does not come directly from the surface potential, but from the potential at the distance of that shear plane. This potential is called the zeta potential ζ. The relationship between the zeta potential ζ and the surface potential Ψ_0 is not defined with certainty. Depending upon the type and the amount of interaction of the ions with each other and with the particle surface and depending upon if the charges originate from acid-base interaction, ion adsorption or the adsorption of polyelectrolytes, the shear plane will be closer or at a farther distance from the surface. The strength of the applied electric field can also play a role[12]. In most cases, the zeta potential will be somewhere between the Stern potential Ψ_S and the eth part (e = Euler number = 2.71828...) of the Stern potential. According to Equation 6.20, $\zeta = \Psi_S/e$ when $\kappa \cdot d = 1$. This is the case[13] when $\kappa = 1/d$. In Figure 6.10, the zeta potential is taken to be close to the Stern potential.

When describing the relationship between the electrophoretic mobility μ and the zeta potential ζ, two cases are distinguished. When the particle diameter is much larger than the extension of the electrostatic double layer, then the Helmholtz-Smoluchowski equation (Equation 6.22) is utilized. This is normally the case in aqueous media.

Equation 6.22 $$\zeta = \frac{\eta}{\varepsilon_r \cdot \varepsilon_0} \cdot \mu$$

When, however, the ratio of particle size to the extension of the double layer is small, then the Debye-Hueckel Equation 6.23 holds.

Equation 6.23 $$\zeta = \frac{3\eta}{2\varepsilon_r \cdot \varepsilon_0} \cdot \mu$$

Equations 6.22 and 6.23 already take the "retardation effects" into consideration. These describe the influence of the ions migrating in the opposite direction to the particles. Contrarily, the constant degradation and restructuring of the outer part of the diffuse double layer ("relaxation effects") is not considered so that the two equations are valid only for zeta potentials up to about 25 *mV*. Equations 6.22 and 6.23 show that the migration of the particles in an electric field is suppressed by the viscosity of the liquid phase, but is directly proportional to

[12] Meaningful measurements are only generated at field strength ranges, at which the electrophoretic mobilities rise linearly with the applied field strength.

[13] For that reason, the inverse of κ, which is defined by Equation 6.21, is called the "inverse Debye length" or the "Debye-Hueckel length".

the dielectric constant of the medium. At the same zeta potential, particles in water ($\varepsilon_r = 80$) would migrate sixteen times faster in an electric field than when suspended, for example, in butyl acetate ($\varepsilon_r = 5$). Because of the small migration speeds, it is hardly possible to perform microelectrophoretic[14] measurements in organic media. From the viewpoint of the author, that is probably the reason why the importance of electrostatic flocculation stabilization in organic media is generally underestimated.

Not only for that reason, but also on account of the ease of the measurements, electro acoustic [27] methods for determining electrophoretic mobilities are becoming more and more popular. In that case, instead of using a DC voltage, an AC (alternating current) voltage with a frequency of about one megahertz is employed. If the pigment particles are charged, then they oscillate in the resulting electrical AC field, generating pressure waves. The amplitude P_{max} of the pressure waves is directly proportional to the electrophoretic high frequency mobility μ_W, the volume fraction of the particles in the suspension φ, the speed of sound within the liquid phase c_S, the density difference between the particles and the liquid phase $\Delta\rho$ and the electric field strength E_W. The proportionality constant K is determined by making a measurement with a calibration solution (for example colloidal silica particles or even salts).

Equation 6.24 $$P_{max} = K \cdot c_s \cdot \varphi \cdot \Delta\rho \cdot E_W \cdot \mu_W$$

One of the two electrodes of the measuring cell is placed on a piezoceramic seismic detector so that the pressure measurements can be performed in real time. With this method, it is not only possible to also determine zeta potentials of pigments in organic media, but also to measure adsorption isotherms of ionic wetting and dispersing agents on pigment surfaces [28, 29]. From the phase shift between the electrical field and the pressure signal, it is also possible to determine the size distributions of the particles.

6.4.5 Electrostatic repulsion energy

When two particles with an electrostatic surface charge meet, then their electrostatic double layers penetrate and a repulsive force comes into being since charges of the same sign repel each other. Unfortunately, there is no distinct, self-contained solution for calculating the force or energy of repulsion that is to be overcome when the particles are brought into each other's neighbourhood. One reason is that surface charges have different causes (see Chapter 6.3.1). Depending upon the type of surface charges, different relationships and boundary conditions prevail. Also, the magnitude of the potential plays a role. In the following, the surface potential

[14] Microelectrophoresis is the determination of the migration speed of charged colloidal particles in an electric field with the aid of a microscope.

is meant to be the effective potential of a particle so that instead of the surface potential Ψ_0, the Stern potential Ψ_S could be meant just as well.

Should the surface charge stem from adsorbed ions, then it must be considered that the ions surrounding the particles ("ion cloud") are mobile and, depending upon their sign, are attracted or repelled by other ions. When two particles carrying a surface charge of the same sign approach, then the concentration of alike ions increases within the gap between them. The potential curves overlap so that the surface charges on the particles increases. The relationship between the surface potential Ψ_0 (or the Stern potential Ψ_S, respectively) and the surface charge density σ_0 (= number of charges per surface unit on the particles) is not linear [30, 31], but, in the case of a 1 : 1 electrolyte, is given by [32]

Equation 6.25 $$\sigma_0 = \sqrt{\frac{2 \cdot \varepsilon_r \varepsilon_0 \cdot c \cdot kT}{\pi}} \cdot sinh\left(\frac{z \cdot e \cdot \Psi_0}{2kT}\right)$$

In Equation 6.22, c is the concentration of ions in the diffuse part of the double layer.

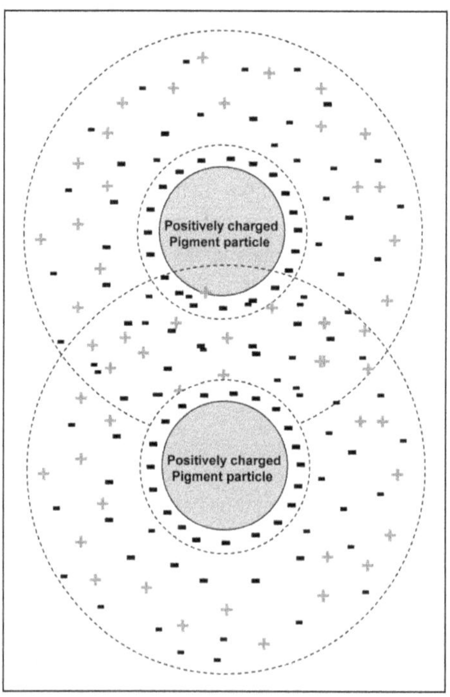

Figure 6.14: Change of ion concentration in the gap between two equally charged particles as they approach

Two separate cases may be distinguished in which either the surface **charge density** or, otherwise, the **surface potential** remains constant during the approach of the particles. Figure 6.14 shows two electrostatically charged particles approaching each other. In doing so, their electrostatic double layers penetrate each other so that the concentration of ions c in the overlapping region increases. This leads to an increase in the value of the square root term of Equation 6.25. So, if the surface potential Ψ_0 remains constant during the approach, then the surface charge density σ_0 must increase. In the case that the surface charge density σ_0 remains constant, then, according to Equation 6.25, the surface potential has to decline, leading to lesser repulsion.

The energy of repulsion between two equally charged pigment particles is not calculated easily. Mathematical approaches lead to equations that cannot be solved analytically. Yet, considering boarder conditions, simplified equations may be deducted. At low surface potentials of up to 50 or 60 mV, the repulsive energy is proportional to the square of the surface potential. Once again, two cases are distinguished, namely that the product between the Debye-Hueckel parameter and the particle radius is either much larger or much smaller than unity. The case $\kappa \cdot r >> 1$ is found in aqueous systems where the ion concentrations can obtain higher values so that κ is larger. Then the electrostatic energy of repulsion may be expressed by

Equation 6.26 $$V_{el} = \frac{1}{2} \cdot \varepsilon_r \varepsilon_0 \cdot r \cdot \Psi_0^2 \cdot \ln(1 + e^{-\kappa d})$$

The case that $\kappa \cdot r << 1$ is met with particles in organic media, where the ion concentrations are low due to low dielectric constants of the liquids. The surface charges lead to potentials that reach very far. The electrostatic energy of repulsion is then better expressed by Equation 6.27.

Equation 6.27 $$V_{el} = \frac{\varepsilon_r \varepsilon_0 \cdot r \; \Psi_0^2}{d + 2r} \cdot e^{-\kappa d}$$

At higher surface potentials, Equation 6.28 is a better approximation [32].

Equation 6.28 $$V_{el} = \frac{64\pi \cdot c \cdot kT \cdot r}{\kappa^2} \cdot \gamma^2 \cdot e^{-\kappa d}$$

The influence of the surface potential is represented in Equation 6.28 by the term γ that is given by Equation 6.29.

Equation 6.29 $$\gamma = \frac{\exp(ze\Psi_0/2kT)-1}{\exp(ze\Psi_0/2kT)+1}$$

Figure 6.15 shows the development of γ^2 as a function of $ze\Psi_0/2kT$. At higher values of the argument in Equation 6.29 the function levels off.

This means that it is not possible to increase the repulsive energy indefinitely by increasing the surface potential. Ultimately, this is the consequence of the exponential development of the surface potential. Extremely high surface potentials primarily come into effect when a very high surface charge density prevails on the particle surfaces.

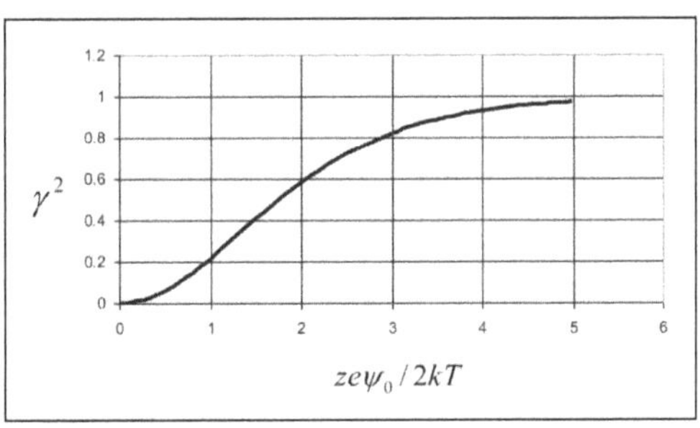

Figure 6.15: Graphic representation of γ^2 (from Equation 6.29)

Yet, the repulsive interaction is generated from potentials at farther distance from the particle surface.

A compilation of the distance dependent repulsive potential curves for a number of different geometric bodies under various conditions is found in [33].

6.5 Steric stabilization

Under certain circumstances, polymer molecules ("macromolecules") adsorbed on pigment surfaces are able to stabilize the pigments against flocculation. This stabilizing mechanism is called "steric stabilization". As is the case with all changes of state in thermodynamics, the free energy of flocculation ΔG_{Floc} is put together by the enthalpy of flocculation ΔH_{Floc} and an entropy term $-T\Delta S_{Floc}$.

Equation 6.30 $$\Delta G_{Floc} = \Delta H_{Floc} - T\Delta S_{Floc}$$

Flocculation takes place if the free energy of flocculation becomes negative, meaning that energy is lost by the system. Contrarily, the system is stabilized against flocculation when ΔG_{Floc} is positive. This may occur under different circumstances.

a) entropic stabilization

In this case, both ΔH as well as ΔS are negative, yet $T\Delta S_{Floc}$, the entropy term in Equation 6.30, is larger in amount than the enthalpy ΔH_{Floc}. This situation is termed "entropic stabilization". Whereas the enthalpy term abets flocculation, the loss in entropy prevents this from happening. In principal, entropically stabilized systems can be brought to flocculate by lowering the temperature.

b) enthalpic stabilization

When both ΔH and ΔS are positive, the system is stabile against flocculation as long as ΔH_{Floc} is larger than the absolute value of $T\Delta S_{Floc}$. In this case, the system is stabilized enthalpically. Whereas the entropy term promotes flocculation, the enthalpy counteracts flocculation of the particles. An increase in temperature may cause these systems to flocculate.

c) mixed enthalpic – entropic stabilization

In the third possible case, both ΔH and ΔS are negative. Both contributions prevent flocculation so that one may speak of a mixed enthalpic-entropic stabilization.

The name "steric stabilization" includes all three cases. As far as the weighting of the different mechanisms is concerned, case a), the entropic stabilization, is probably the most frequent, followed by case c), the mixed enthalpic-entropic stabilization.

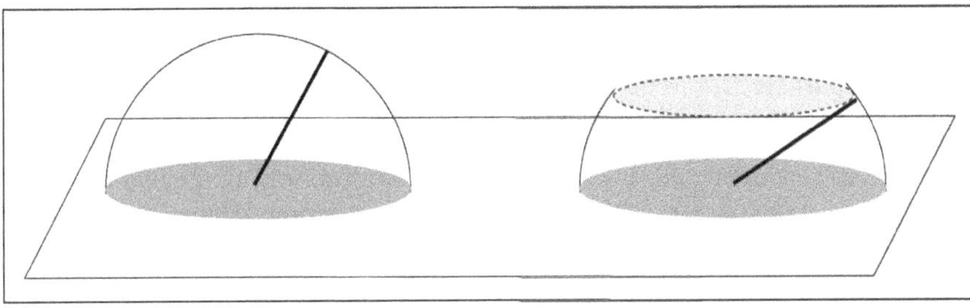

Figure 6.16: Model of Mackor[35, 36] *for explaining the loss of entropy by constrained volume acccess of adsorbed rod shaped molecules on pigment surfaces (right side)*

There is reason to assume that the entropy ultimately plays the major role in view of stabilization against flocculation. *Napper*[34] describes a system consisting of polyvinylacetate dispersed in water which changes from enthalpic to entropic stabilization at the addition of dioxane.

Different models were proposed to explain the influence of entropy loss on flocculation stabilization.

Mackor and *van der Waals*[35, 36] used carbon black in C-16 paraffin as a model for entropic stabilization. They considered the paraffin molecules as rods that adsorbed on to the surface of the carbon black pigments with one end (see Figure 6.16). With the exception of this attached point, the rods are free to take any position so that the opposite ends of the rods can be anywhere on a hemisphere. When a second particle approaches to a distance that is shorter than the rod length, the number of positions that the attached rods may take are restrained. Their entropy diminishes so that, even at negative enthalpies, flocculation may be prevented according to Equation 6.30.

A refined model comes from *Clayfield* and *Lumb*[37, 38]. The authors calculated the entropies of a polymer chain growing from a model pigment surface in a cubic lattice with a Monte Carlo statistic by determining the number of possible configurations (states) and found the entropy loss when these were reduced. They compared the repulsion generated from the entropy loss with the van der Waals attraction of the particles and calculated maxima and minima of the potential curves. Their conclusions were:

- An incomplete surface coverage of the particles reduces stabilization
- Flexible polymer chains promote stabilization since the entropy loss upon flocculation is higher
- As expected, higher Hamaker constants of the particles enhance flocculation
- The thicker the polymer layers are in comparison to the particles sizes, the lower the energy barriers are and the lower the secondary minima become

Their calculations also indicate that, for a given polymer, there is a minimum particle size under which entropic stabilization can no longer take place. The authors expect this particle size to be in the order of approximately 10 nm.

However, the model of Clayfield and Lumb describes a polymer/solvent system in which the solvent molecules have no additional interaction with the non-adsorbed part of the macromolecules. An example of such a system is considered to be a block copolymer with an anchor block that is pigment affine and a second block extending into the solution, consisting of a polyolefine. A hydrocarbon acts as a solvent. This is a description of a so-called "theta system" (see below). The model of Clayfield and Lumb is therefore based on the assumption of an elastic compression of the adsorbed polymer molecules on the event of particle to particle approach.

Contrarily, *Fischer* [39] ascribed the stabilizing properties of adsorbed polymer molecules to the interactions between the polymer and the solvent molecules. He utilized the results of the Flory-Huggins theory on polymer solutions. Later, *Napper* [40–42] continued this line of thought and verified it experimentally.

6.5.1 Macromolecules in solution

Polymers dissolve in solvents when their interaction with the solvent molecules is more pronounced than the interaction between the polymer segments themselves. When, for example, a solid block of poly vinyl chloride (PVC) is placed into a dish filled with acetone, the polymer first begins to swell. The volume of the block increases continuously until, finally, the polymer goes into solution. PVC consists of a hydrocarbon chain on which at every second carbon atom a hydrogen atom is replaced by chlorine.

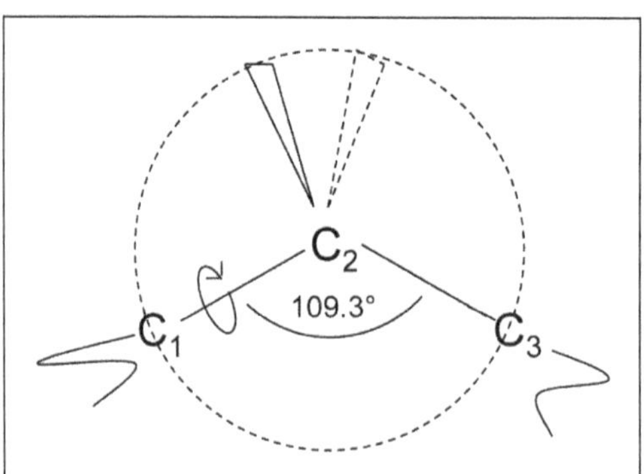

Figure 6.17: sp^3-hybridized hydrocarbon. All angles between the chemical bonds that extend from the carbon atom C_2 have an angle of 109.3 °(tetrahedron angle).

In PVC the carbon atoms are "sp^3-hybridized". The chemical bonds that extend from each carbon atom are pointed into the four corners of a tetrahedron (triangular pyramid; see Figure 6.17). For that reason the main chain of a polymer backbone is not elongated, but every C-C-C structure element has an angle of (approximately) 109.3° instead (=tetrahedron angle).

In addition, all atoms along a single bond are free to rotate. When a hypothetical polymer chain is allowed to grow in a computer simulation, then every carbon atom that is added can either lay to the left, the right, above or below and in front of or in the back of the carbon atom that it is connected to. Such a mathematical formalism is called "random walk[15]" [43]. Figure 6.18 shows a section of such a growing chain in a schematic way. After many thousands of growing steps, a round or elliptical (bean-shaped) body is formed which has a low density due to its large void volume.

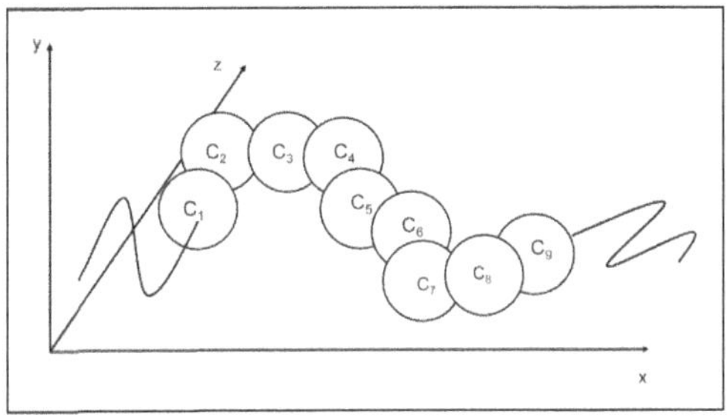

Figure 6.18: Three dimensional extension of a of a growing carbon chain

In the process, the segment density decreases within an ellipsoid from the center to the outskirts. This is schematically shown in Figure 6.19.

When a macromolecule is in solution, the "void volume" is actually filled with solvent. When the interaction between the solvent molecules and the individual polymer segments (monomer units) are just as large as the interaction forces between the polymer segments themselves, then the polymer chains are suspended "stress free", i.e., all the C-C-C bonds have an angle of 109.3° (see Figure 6.19 b). This case is named the θ-("theta") condition. If the interaction between the polymer segments and the solvent is larger, then the carbon chain is expanded. The C-C-C bond is then larger than the tetrahedron angle and the polymer molecules has a larger "hydrodynamic volume" (see Figure 16.9 a). If, however, the interaction

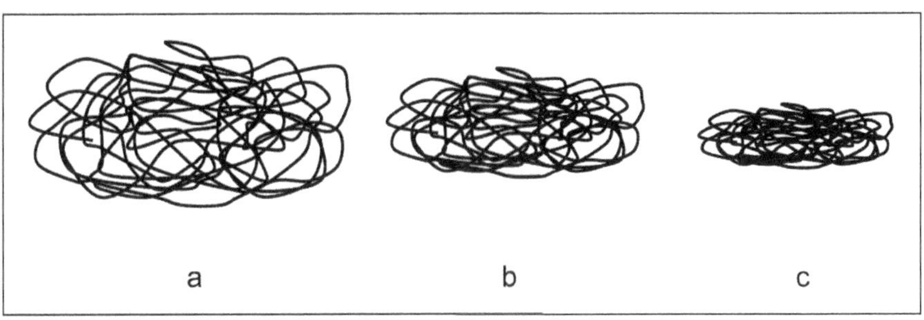

Figure 6.19: Conformation of a polymer in solution.
a) > Theta condition; C-C-C bond angle > 109.3°; $\chi < 0,5$
b) = Theta condition; C-C-C bond angle = 109.3°; $\chi = 0,5$
c) < Theta condition; C-C-C bond angle < 109.3°; $\chi > 0,5$

[15] There are nice computer simulations of random walks available on the internet.

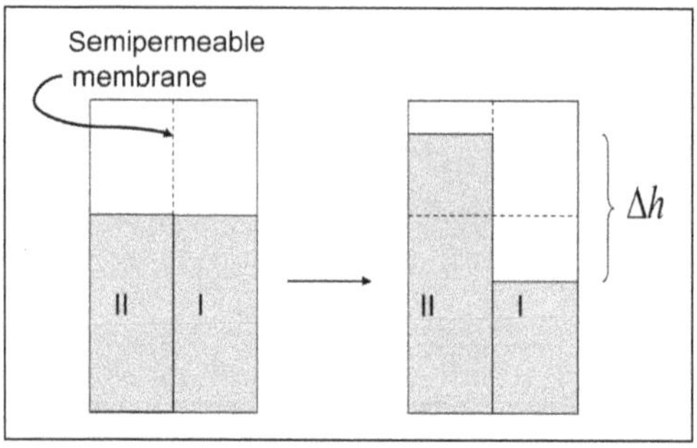

Figure 6.20: Principle of osmometric measurements
I: Chamber filled with pure solvent
II: Chamber filled with polymer solution
Δh: Difference in height between the surface of the solvent and the polymer solution after equilibrium is reached.

between the polymer segments and the solvent is less than the interaction between the polymer segments themselves, then the macromolecules fill a smaller volume. The C-C-C bond angle is smaller than 109.3°. This case is shown on the right side in Figure 6.19 c.

When, starting from the case of a very good solvent (Figure 6.19a), a thermodynamically poor solvent is added to the solution, then the polymer-solvent interactions are reduced and the polymer solution will change into the direction of the θ-state. In principle, the same change in polymer swelling can also be achieved by changing the temperature (in most cases by lowering it). As soon as the θ-state is reached, the solvent-polymer interactions compete with the interactions between the polymers themselves. Under θ-conditions, the polymers are just barely dissolved. Yet, as soon as the conditions are less than θ, the interaction amongst the polymer segments prevail and the macromolecules that meet due to Brownian motion coagulate and drop out.

Interactions between solvents and polymers can be studied with the help of osmometric measurements. In osmometry, solutions of polymers are separated from a pure solvent by a membrane that is permeable only for solvent molecules ("semi-permeable membrane"). If the solvent molecules undergo more interaction with the polymers than amongst themselves, then an "osmotic pressure" is generated so that the solvent molecules flow through the membrane into the chamber holding the polymer solution[16]. Figure 6.20 explains the principle. The mass transport leads to a difference in the heights of the liquids in the two chambers that correspond to a hydrostatic pressure Π. In the case of a θ-solvent, the van't Hoff equation holds whereby the quotient between Π and the polymer concentration at the beginning of the measurement c is inversely proportional to the number average of the polymer's molecular weight M_n.

Equation 6.31 $$\frac{\Pi}{c} = RT \cdot \frac{1}{M_n}$$

[16] In osmometry, the polymer-solvent system acts just like in the experiment explained formerly, when a block of solid polymer is place in pure solvent, only that polymer and solvent are both separated by a semi-permeable membrane.

When Π is measured for different polymer concentrations and $\Pi/(cRT)$ is plotted against c, then a straight line is obtained with an intercept with the y-axis at $1/M_n$ and a slope of zero. If the polymer solution is not in the θ-state, then either an ascending (if better than θ-condition) or a descending (if less than θ-condition) line or, otherwise, a curve instead of a straight line is found. In that case, the expanded van't Hoff equation is applicable.

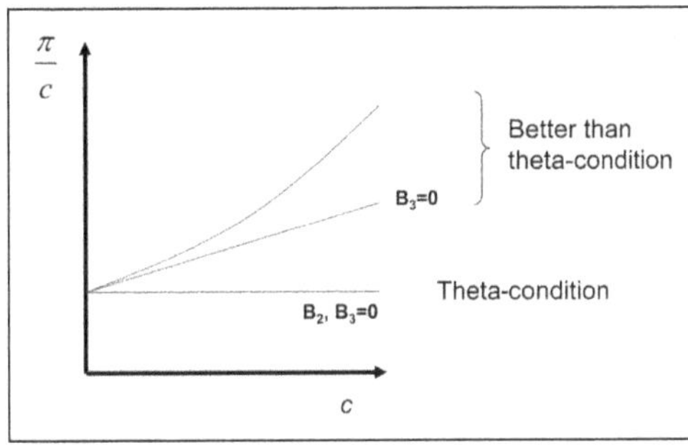

Figure 6.21: Examples for osmometric measurements

Equation 6.32 $$\frac{\Pi}{c} = RT\left[\frac{1}{M_n} + B_2 \cdot c + B_3 \cdot c^2 + ...\right]$$

B_2 and B_3 are called the second and third virial coefficient, respectively. They can be found by fitting Equation 6.32 to measured values of Π/c. When B_3 happens to be zero, a plot of the reduced osmotic pressure vs. polymer concentration yields a straight line with the slope B_2. Figure 6.21 depicts the different possibilities. An example of a theta system is nitrocellulose at room temperature in nitro benzene, whereas methanol and acetone, for example, are thermodynamically very good solvents at ambient temperature for nitrocellulose and lead to better than θ-conditions.

According to the Flory-Huggins theory, [44-49], the second virial coefficient depends upon V_1, the molar volume of the solvent and V_2, the molar volume of the polymer, as well as on its molecular weight and the so-called interaction parameter χ (Greek alphabet letter "Chi"). χ is also called the Flory-Huggins interaction parameter.

Equation 6.33 $$\frac{\Pi}{c} = RT\left[\frac{1}{M_n} + \frac{v_2^2}{V_1} \cdot (0{,}5 - \chi) \cdot c + ...\right]$$

v_2, the partial specific volume of the polymer, is given by the quotient between the molar volume and the molecular weight of the polymer ($v_2 = V_2/M_n$). A comparison of Equation 6.33 with Equation 6.32 leads to the connection between the second virial coefficient and the interaction parameter

Equation 6.34 $$B_2 = \frac{v_2^2}{V_1}\,(0{,}5 - \chi)$$

The smaller the Flory-Huggins interaction parameter is, the more interaction occurs between the solvent molecules and the polymer (see Figure 6.22). $\chi < 0.5$

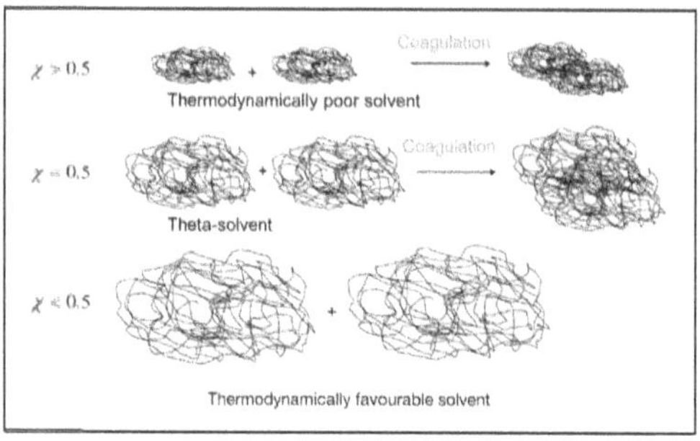

Figure 6.22: Correlation between the swelling behavior of a solvent and its ability to dissolve a polymer

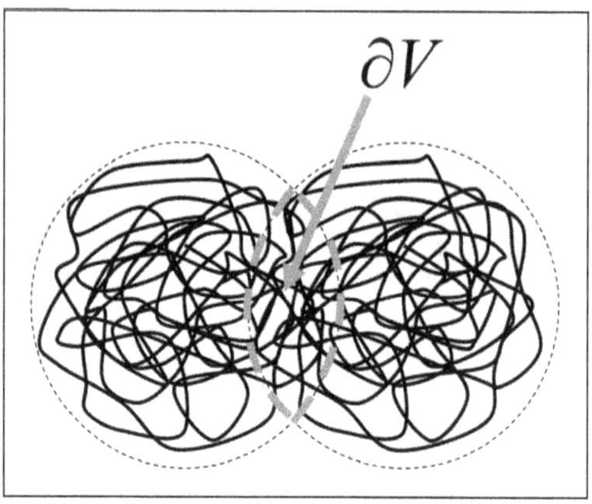

Figure 6.23: Overlap of the volumes of two polymer molecules in solution when penetrating each other.

means that the solvent is thermodynamically favourable so that the polymer is readily dissolved. When $\chi = 0.5$, the solvent evokes theta conditions. Resin coagulation may begin. When $\chi > 0.5$, the second virial coefficient becomes negative. The polymer is no longer soluble but coagulates and drops out. Sometimes in English literature the interaction parameter is called "solvent power[17]". The Flory Huggins interaction parameter effects the "conformation", that is the shape and the hydrodynamic volume of the dissolved polymers as shown in Figures 6.22 and 6.19.

As already mentioned, the dissolution of a polymer in a solvent is also temperature dependent, although this is not accounted for in Equation 6.33. Normally the solubility of a polymer is increased at higher temperatures. *Flory* and *Krigbaum* [50] extended the Flory-Huggins theory of the solubility of macromolecules by defining the interaction parameter χ as the sum of an enthalpic (index H) and an entropic (index S) contribution (χ_H and χ_S), whereby the enthalpic term is temperature dependent.

Equation 6.35 $$\chi = \chi_S + \chi_H$$

They studied the case of two dissolved polymer molecules (indexes l and m) meeting and penetrating each other. Figure 6.23 shows the underlying model. In order to describe the changes in the enthalpy $\partial \Delta H$ and the entropy $\partial \Delta S$, they introduced an enthalpy parameter κ_l and an entropy parameter Ψ_l.

[17] Unfortunately, both expressions are misleading since the interaction between polymer and solvent is more pronounced, the **smaller** the value of the interaction parameter is!

Equation 6.36 $$\partial \Delta S = -2kT \cdot \Psi_1 \cdot \frac{v_2^2}{V_1} \cdot \rho_l \rho_m \cdot \partial V$$

Equation 6.37 $$\partial \Delta H = -2kT \cdot \kappa_1 \cdot \frac{v_2^2}{V_1} \cdot \rho_l \rho_m \cdot \partial V$$

In Equations 6.36 and 6.37, ρ_l and ρ_m are the polymer segment densities of the two polymers in the overlap region ∂V. Due to the negative signs in Equations 6.36 and 6.37, negative values of the entropy parameter and enthalpy parameter correlate with an increase in entropy or enthalpy, respectively, when the polymers intermingle. According to Flory and Krigbaum,

Equation 6.38 $$(0.5 - \chi) \equiv (0.5 - \chi_S - \chi_H) \equiv \Psi_1 - \kappa_1 \equiv \Psi_1 \left(1 - \frac{\theta}{T}\right)$$

whereby $\Psi_1 = 0.5 - \chi_S$ and $\kappa_1 = \chi_H$.

Replacing $(0.5 - \chi)$ of Equation 6.33 by $\Psi_1(1 - \theta/T)$ leads to

Equation 6.39 $$\frac{\Pi}{c} = RT\left[\frac{1}{M_n} + \frac{v_2^2}{V_1} \cdot \Psi_1 \cdot \left(1 - \frac{\theta}{T}\right) \cdot c + ...\right]$$

In Equation 6.39, θ is the temperature at which theta conditions are reached. Therefore, the expression "theta temperature" is sometimes used synonymously for theta conditions. Equation 6.38 discloses that $\chi = 0{,}5$ when $\theta = T$. Then, both sides of the equation turn to zero.

The entropy parameter can be determined by performing of osmometric measurements at different temperatures in case the theta temperature is known or, otherwise, is measured separately. For that, the value of the second virial coefficient is plotted against the temperature and is extrapolated to a value B_2 of zero. The corresponding temperature is the theta temperature θ. Subsequently, the entropy parameter is found from the slope of a reduced pressure Π/c versus concentration c plot. Finally, from the third and fourth term of Equation 6.38, the enthalpy parameter κ_1 can be calculated using the expression $\kappa_1 = \Psi_1 \cdot \theta/T$.

The significance or the Flory-Krigbaum theory lies in the fact, that it offers a conclusive description of the parameters that influence the solubility of polymers in solvents and that these parameters may be ascertained by simple osmometric measurements. In addition and as pointed out in Chapter 6.5, these theories are useful for explaining the affect of adsorbed polymer molecules on the stabilization of pigments against flocculation. This will be explained further in the following section.

6.5.2 Macromolecules on pigment surfaces

The precondition for steric stabilization of pigments and extenders in polymer solutions is, first of all, that the molecules adsorb to the surfaces of the particles. As a general rule, those parts of the molecules having the lowest solubility in the liquid phase are adsorbed the most. This principle is the basis for all kinds of liquid chromatography separations and is, as such, well known. Exceptions are encountered when molecules are able to chemisorb onto the surfaces. In that case, the effect of a possibly good solubility of a molecule is overridden by the formation of chemical bonds with surface moieties and the enthalpy connected to that reaction. Chemisorption, for example, occurs when molecules with carboxyl groups (-COOH) react with titanium dioxide surfaces. Phosphates, as another example, chemisorb to aluminium oxide particles as well as to titanium dioxide surfaces.

When a dissolved, linear polymer molecule meets a pigment surface, then single polymer segments may adsorb. Since the polymer constantly changes its conformation, by and by also other parts of the polymer may change to an adsorbed state. In the course of time (seconds to hours), the conformation of the polymers change in a way that is shown schematically in Figure 6.24. Next to "trains" that anchor the polymer to the pigment surface, "loops" and "tails" are formed that reach out into solution. A branched or cross-linked polymer, which might show less ability to orientate differently, will change its conformation less owing to its restrained possibilities.

The adsorption model underlying Figure 6.24 originally stems from *Jenckel* and *Rumbach* [51]. The coherence of the model has been verified by many studies, inter alia by IR-spectroscopic investigations [52, 53].

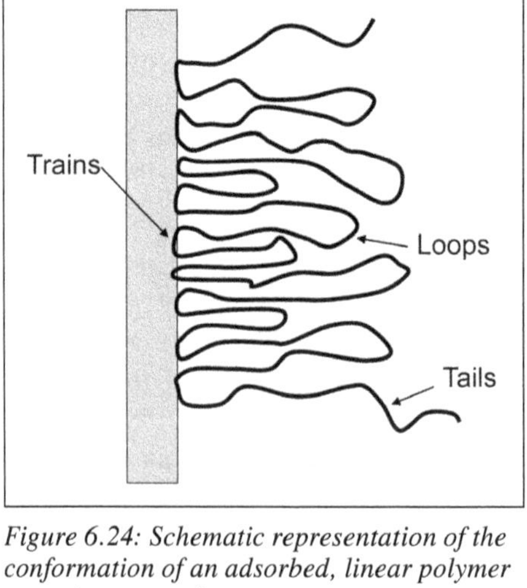

Figure 6.24: Schematic representation of the conformation of an adsorbed, linear polymer on a pigment surface

Under normal circumstances, adsorbed polymers are more or less stable against desorption. In order to completely separate from a surface, all of the trains have to disengage. This becomes less probable, the higher the molecular weight of the polymer is. That is why a number of authors have found that, at equilibrium, the higher molecular weight fractions of a polymer are preferentially adsorbed [54-58]. The changes in the molecular weight distribution before and after adsorption are easily monitored using gel permeation chromatography techniques.

Figure 6.24 also demonstrates that the segment density distribution of the adsorbed polymer changes with the distance from the surface. The highest segment density is found directly on the pigment surface, gradually falling to a value of zero at the distance where the polymer reaches farthest into the solution. According to *Fischer* [39] and *Napper* [40-42], the stabilizing action of adsorbed polymers can be explained, analogously to the Flory-Huggins theory on the solubility of polymers, by the effect of an increase in polymer concentration in the adsorption layer when particles approach each other. Figure 6.25 represents sections of two pigment surfaces, each covered with adsorbed polymer molecules. When the particles come near each other, the two opposing adsorption layers penetrate so that the polymer concentration in the overlapping region is increased.

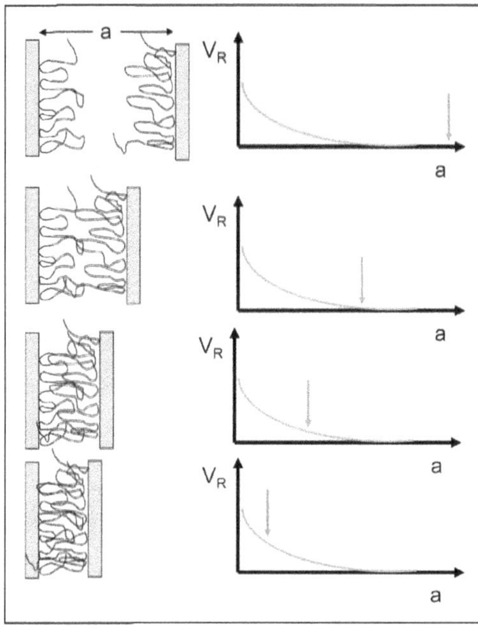

Figure 6.25: Increase of polymer concentration in the gap between two particles with adsorbed polymer layers and its influence on the repulsive energy

The total energy of repulsion V_R is given by the sum of the repulsive energy contributions of all the polymers that overlap during the encounter of the particles.

When the particles are separated at a distance a from one another, the integral of overlap S_a of the polymer molecules is given by

Equation 6.40 $$S_a = \int_V \rho_i \cdot \rho_k \cdot dV$$

The overlap leads to an osmotic pressure of the solvent so that it streams into the gap between the particles where the polymer concentration is higher than elsewhere in the system. This lets the particles separate from one another again. The energy of repulsion is therefore

Equation 6.41 $$V_R = 2kT \cdot \frac{v_2^2}{V_1} \cdot (0{,}5 - \chi) \cdot S_a$$

or otherwise, following Equation 6.35,

Equation 6.42 $$V_R = 2kT \cdot \frac{v_2^2}{V_1} \cdot \Psi_1 \left(1 - \frac{\theta}{T}\right) \cdot S_a$$

For two spherical particles with a radius r and an adsorbed polymer layer with a thickness d and a polymer segment density of $\rho_l = \rho_m = \rho$, the overlap integral (Equation 6.40) becomes [39]

Equation 6.43 $$S_a = \rho^2 \cdot \frac{2}{3}\pi \cdot \left(d - \frac{a}{2}\right)^2 \left(3r + 2d + \frac{a}{2}\right)$$

Insertion of Equation 6.43 into Equation 6.41 and 6.42 leads to manageable expressions for the repulsive energies of sterically stabilized pigment particles

Equation 6.44 $$V_R = \frac{4}{3}\pi \cdot kT \cdot \frac{v_2^2}{V_1}(0{,}5 - \chi) \cdot \rho^2 \cdot \left(d - \frac{a}{2}\right)^2 \left(3r + 2d + \frac{a}{2}\right)$$

Equation 6.45 $$V_R = \frac{4}{3}\pi \cdot kT \cdot \frac{v_2^2}{V_1} \cdot \Psi_1 \left(1 - \frac{\theta}{T}\right) \cdot \rho^2 \cdot \left(d - \frac{a}{2}\right)^2 \left(3r + 2d + \frac{a}{2}\right)$$

Just like in the case of electrostatic stabilization, it is hardly possible to make quantitative forecasts on the stabilization of pigment particles by steric mechanisms. A fundamental precondition is, naturally, that the polymers are adsorbed to the particle surface. For that to happen, their solubility should not be extensively well. Furthermore, the coverage of the surface should be at about 100 %, i.e., the particle surface should be completely protected. As far as the adsorption layer itself is concerned, Equations 6.44 and 6.45 clarify some fundamental relationships.

- Repulsion only takes place when the entropy parameter ψ_1 has a positive sign, or if the Flory-Huggins interaction parameter χ is less than 0.5. A positive value for ψ_1 means that the entropy becomes less during flocculation. In that case, the repulsive energy at a certain distance becomes larger at higher temperatures. Flocculation is promoted, however, when ψ_1 has a negative value.
- The higher the segment density of the polymers in the adsorption layer is, the higher the repulsive energy is expected to be, provided that ψ_1 is positive and χ is smaller than 0.5.
- Thicker adsorption layers let the onset of repulsion occur at larger particle-particle distances.
- The larger the particles are, the larger the overlap volume of the adsorbed polymer layers will be when the particles approach each other and the larger the repulsive energy will be also (see Figure 6.26).

Steric stabilization will not take place in the absence of a solvent.

The last statement is to a certain sense trivial, yet momentous. It explains why pigment performance, for example the tinting strength of coloured pigments or the lightening power of white pigments, is normally poorer in plastic melts and powder coatings than in formulations containing dissolved polymer molecules. It also gives the reason why pigments designed for use in plastics are often treated with poly dimethyl siloxanes. These organic post treatments on the one hand reduce the mechanical strengths of the pigment agglomerates, but on the other hand hinder the adsorption of polymer molecules that might possibly stabilize by a steric mechanism. Yet, since there are no solvents in the system, which is a prerequisite for steric stabilization, in the end, the advantage of easy deagglomeration of thus treated pigments prevails.

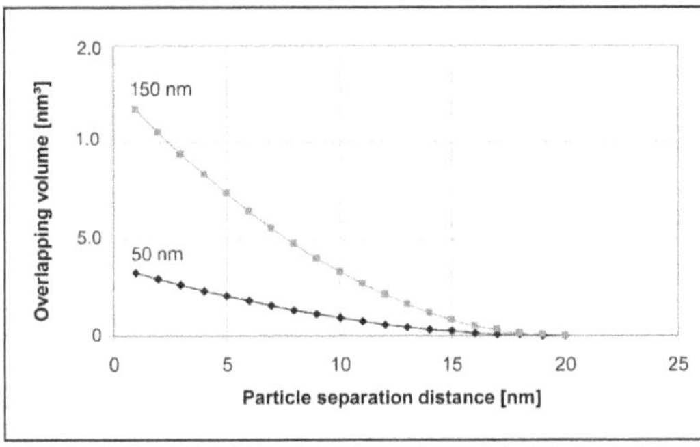

Figure 6.26: The overlap integral S_a (Equation 6.43) of two particles with a diameter of 50 nm or 150 nm, respectively, and an adsorption layer of 10 nm thickness

Another property of polymer solutions that is strongly dependent upon the second virial coefficient is the viscosity. For ecological as well as economic reasons, it is always a goal to keep the solid contents (content of non-volatile components) of paint formulations as high as possible. That is why medium solid and high solid resins were developed. The ability of these formulations to flow freely is provided for on the one hand by low molecular weights of the resin molecules and, on the other hand, by the use of thermodynamically poor solvents so that the hydrodynamic volumes of the macromolecules remain as small as possible. Both measures lead to a lower resistance to flow and therefore low viscosity even at high solid contents. In the view of pigment stabilization, these are not ideal conditions since, although the adsorption of resin happens to a high degree, the adsorption layer thicknesses are weak and, due to high χ-values, they are compressed. For that reason, the osmotic pressure during flocculation is less pronounced, leading to the well-known problems that are encountered particularly in these systems. As expected, in a study of various medium and high solid resins with molecular weights down to 1000 Dalton and second virial coefficients down to 10^{-4} $mol \cdot cm^3/g$, it was found that pigment stabilization became poorer the lower the molecular weight and the lower the second virial coefficient was [59]. Normally, in these cases only high molecular weight dispersion additives will remedy the situation. Effective dispersion additives for steric stabilization normally include segments of low solubility that act as anchor groups and, on the other hand, parts that have a lot of interaction with organic solvents. The former groups establish the trains whereas the latter ones protrude into the solution as loops and tails (see Figure 6.24) and ensure stabilization. Molecular architecture is very versatile and is by no means confined to linear copolymers or block copolymers, respectively. Many of these products have a comb-shape with a pigment affine backbone and stabilizing side chains.

As a rule, an even distribution of the pigment particles is aspired in polymeric composite materials. Still, one exception is when antistatic pigments are used in paints. Examples for antistatic paints are primers for plastic parts that are further covered with electrostatically applied top coats or, otherwise, floor paints in explosion proof areas or cleanrooms. In these formulations, the antistatic pigments must touch each other in the cured paint film so that the composites can discharge electrostatical loading. The pigment volume concentration (PVC) that is necessary for this to happen is called the "percolation PVC". If the pigment particles are well dispersed and evenly distributed, much pigment is necessary to achieve antistatic performance. In sterically stabilized systems, it is possible to induce flocculation by adding thermodynamically poor solvents. When this happens, a network or band like structure of flocculated pigments incurs that can act as conducting paths next to areas that are free of pigments. The PVC of costly antistatic pigments may be reduced in this way by up to 40 % [60].

Steric stabilization of pigments can occur in solvent based systems as well as in aqueous formulations. Contrary to electrostatic stabilization, it is independent of the prevailing ionic strength in solution. When polymeric salts like polyacrylates adsorb onto pigments in aqueous surroundings, stabilization is caused by a mixture of electrostatic repulsion and the steric mechanism.

6.6 Solubility parameters

The Flory-Huggins interaction parameter describes the interaction only between a certain polymer and a particular solvent. A totally different approach is to assign measures to the solvents that relate to their overall ability to dissolve any other matter, the so called "solubility parameters".

The dissolution of a polymer in a solvent can be understood as a mixing process. For mixing reactions also, the Gibbs equation can be set up, whereby the free energy of mixing ΔG_m depends upon the mixing enthalpy ΔH_m and the mixing entropy ΔS_m.

Equation 6.46 $$\Delta G_m = \Delta H_m - T\Delta S_m$$

The mixing of two substances takes place if the free energy of mixing is negative. Since the entropy in mixtures is generally raised, the entropy term as a rule favours the mixing process. Therefore, the mixing enthalpy becomes crucial for the mixing to take place or not. When heat is generated while mixing, i.e. if mixing is exothermal, then the enthalpy also promotes mixing. In the case of endothermal mixing enthalpies, mixing can either not take place or may happen only when the system is heated.

When two liquids A and B are brought together, then the interaction energies $E_{A\text{-}A}$ and $E_{B\text{-}B}$ of the pure liquids are exchanged for the interaction energy $E_{A\text{-}B}$ of the mixture.

$$E_{A-A} + E_{B-B} \rightarrow 2E_{A-B}$$

If $E_{A\text{-}A}$ is much larger or much smaller than $E_{B\text{-}B}$, then mixing will not occur. In the first case, too much energy would be necessary to overcome the interaction between the molecules of the type A. In the second case, the interaction between the molecules of type B prevent mixing to happen. This simple contemplation explains the reason for the assertion of medieval alchemists, that similar substances will dissolve similar substances (Latin: Similia similibus solvuntur).

A measure for the interaction energy between substances is their evaporation enthalpy ΔH_{vapor} (enthalpy of vaporization). Upon evaporation of a liquid, part of the evaporation enthalpy is needed to separate the molecules from one another. This part is the energy of evaporation ΔE_{vapor}. Another part is necessary to then expand the volume, that is to distribute the molecules within the volume V. The work of volume expansion is given by the product between the pressure P and the volume V which, according to the ideal gas law, is equal to the product of the gas constant R and the absolute temperature T in Kelvin (Equation 6.47 relates to one mole of gas).

Equation 6.47 $$P \cdot V = R \cdot T$$

So,

Equation 6.48 $$\Delta H_{vapor} = \Delta E_{vapor} + RT$$

Or

Equation 6.49 $$\Delta E_{vapor} = \Delta H_{vapor} - RT$$

When the energy of vaporization is standardized to the volume V of the liquid, the volume based evaporation energy is found which is also called the "cohesive energy density". The square root of the cohesive energy density is named the solubility parameter δ. Thus, the solubility parameter of a liquid is

Equation 6.50 $$\delta = \left(\frac{\Delta E_{vapor}}{V}\right)^{1/2} = \left(\frac{\Delta H_{vapor} - RT}{V}\right)^{1/2}$$

δ is called the "Hildebrand solubility parameter". Convenient number values arise if the solubility parameters are reported[18] in the unit $(J/cm^3)^{1/2}$.

[18] $(J/cm^3)^{1/2} = 10^6(J/m^3)^{1/2} = 0.4887\ (cal/cm^3)^{1/2} = (MPa)^{1/2}$. In older literature, solubility parameters have the dimension $(cal/cm^3)^{1/2}$.

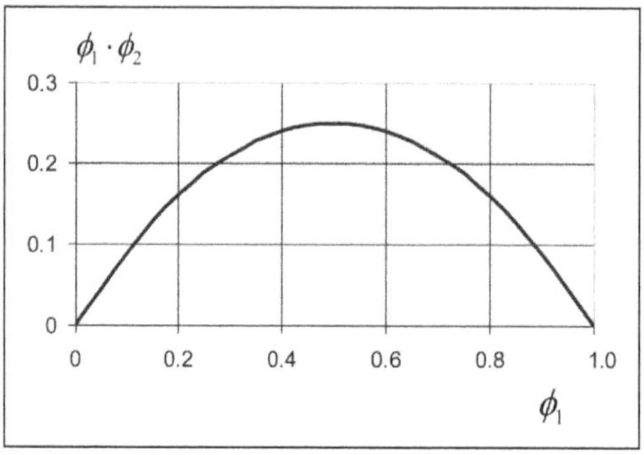

Figure 6.27: Presentation of the product of two volume fractions as a function of the volume fraction of substance 1. $\phi_1 + \phi_2 = 1$

According to *Hildebrand* and *Scott* [61] as well as *Scatchard* [62], the mixing enthalpy for two liquids 1 and 2 with the total volume V_{tot} is given by

Equation 6.51

$$\Delta H_m = V_{tot}(\delta_1 - \delta_2)^2 \cdot \phi_1\phi_2$$

whereby ϕ_1 and ϕ_2 are the volume fractions of the two liquids. The term $\phi_1 \cdot \phi_2$ is presented graphically in Figure 6.27. It is zero if either ϕ_1 or ϕ_2 becomes zero and has a maximum of 0.25 at $\phi_1 = \phi_2$. $(\delta_1 - \delta_2)^2$ is an upward opened parabola with a minimum of zero at $\delta_1 - \delta_2 = 0$. As a rule, two liquids will mix readily if their solubility parameters differ by no more than $\Delta\delta \leq 4 - 6\ (J/cm^3)^{1/2}$.

For molecules that undergo dispersive interactions exclusively, the Hildebrand solubility parameter δ is a good measure. If, however, polar interactions and hydrogen bridges supervene, then the Hildbrand solubility parameters are no longer sufficient as a sole criterion. A number of different proposals were made to consider of polar interactions also. One of these comes from *Hansen* [63-65] who splits the solubility parameter into different contributions from dispersive interactions (index d), dipole-dipole interactions (index p) and hydrogen bonds (index h) in such a way that

Equation 6.52 $$\delta^2 = \delta_d^2 + \delta_p^2 + \delta_h^2$$

There are different ways to determine the individual contributions. Following a proposal of *Bondi* and *Simken* [66], *Hansen* utilized the enthalpies of evaporation of "homomorph" molecules that undergo only dispersive interactions for determining δ_d. Homomorph molecules differ from the actual molecule only in their lacking the polar entity. The homomorph of butanol, for example, is butane. The difference in the square of the Hildebrand solubility parameter and the square of the dispersive part of the solubility parameter yields the polar contribution of the solubility parameter δ_a.

Equation 6.53 $$\delta_a^2 = \delta^2 - \delta_d^2 = \delta_p^2 + \delta_h^2$$

In quite tedious experiments, Hansen studied the solubility of polymers in a large number of different solvents to find the different contributions to δ_a in a semi-empirical way. Later on he refined the solubility parameter contributions by introducing corrections for specific functional groups in the solvent molecules. There are a number of semi-empirical relationships between the three-dimensional solubility parameters and other quantities. According to [67-68], the dispersive part of the solubility parameter may be calculated from the refractive indexes of the solvents, for example.

Equation 6.54 $$\delta_d = 9.5 \cdot n_D - 5.55$$

For determining the polar part of the solubility parameter, *Hansen* and *Skaarup* [65] used a relation of *Boettcher* [69], into which the dipole moment μ, the (static) relative dielectric constant ε_r and the molar volume V_m enter.

Equation 6.55 $$\delta_p^2 = \frac{50694}{V_m^2} \cdot \frac{\varepsilon_r - 1}{2\varepsilon_r + n_D^2} \cdot (n_D^2 + 2)\,\mu^2$$

δ_P is found in the physical dimension $(J/cm^3)^{1/2}$ if Equation 6.55 is employed. A further formula that is also reported to generate meaningful results [70] is from *Beerbower* [71].

Equation 6.56 $$\delta_p^2 = 9.5 \cdot \frac{\mu}{\sqrt{V_m}}$$

In order to estimate the contribution of N OH-bonds of a molecule, Hansen and Skaarup suggested to use the relationship [65]

Equation 6.57 $$\delta_h = \sqrt{\frac{20934 \cdot N}{V_m}}$$

In this case too, δ_h has the physical dimension $(J/cm^3)^{1/2}$.

Equation 6.52 shows that the solubility parameter of Hildebrand can be understood as a vector of the length δ in a three dimensional space that is spanned by the orthogonal (standing perpendicular to each other) vectors δ_d, δ_p and δ_h. This is illustrated in Figure 6.28. In order for two liquids to be miscible, it is not sufficient for the Hildebrand solubility parameters to be similar (i.e. the vectors δ have the same length), but the different contributions should be alike also. In other words, the vectors in the vector space should be close to each other.

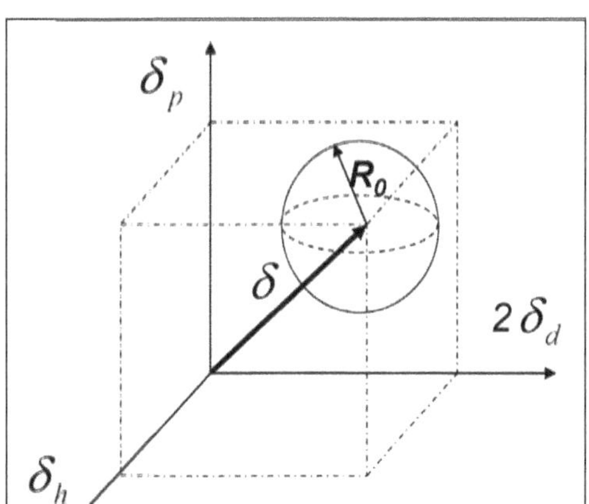

Figure 6.28: Three-dimensional representation of the solubility parameters according to Hansen

In their experiments on the solubility of polymers, Hansen and Skaarup found that the solvents dissolved the macromolecules as long as the distance of the vectors did not exceed a certain limit. More than that, they found that the solubility limits could be visualized as a spherical space if the δ_d-axis was drawn with a scale twice as long as δ_h and δ_p. The solvency limit is then defined by a sphere of radius R_0, whose centre is given by the solubility parameters of the polymer (see Figure 6.28). At the solubility limit, the free energy of mixing ΔG_m is zero. If $\Delta\delta_d$, $\Delta\delta_p$ and $\Delta\delta_h$ are the differences of the solubility parameters between polymer and solvent, then the distance R in the solubility parameter vector space is

Equation 6.58 $$R = \sqrt{4 \cdot \Delta\delta_d^2 + \Delta\delta_p^2 + \Delta\delta_h^2}$$

According to the assertions given above, the polymer will be dissolved if $R < R_0$.

It is of great importance to note that in mixtures, the solubility parameters interact according to the principle of linear momentum. The effective solubility parameter is calculated from the sum of the volume fractions of the solvents in combination with their single solubility parameter values.

Table 6.1: Solubility parameters of a number of solvents according to Hansen

source: Shell Chemie Brochure "Löslichkeitsparameter, (= solubility parameter)", LA-08

Solvent type	Compound	δ_d	δ_p	δ_h
Aliphatic carbohydrates	n-Hexane	7.3	0.0	0.0
	Cyclohexane	8.2	0.0	0.0
Aromatic carbohydrates	Toluene	8.8	0.7	1.0
	o-Xylene	8.7	0.5	1.5
	Ethylbenzene	8.7	0.3	0.7
Ketones	Methyl ethyl ketone	7.8	4.4	2.5
	Methyl isobutyl ketone	7.5	3.0	2.0
	Isophorone	8.1	4.0	3.6
Esters	Ethylacetate	7.7	2.6	3.5
	n-Butylacetate	7.7	1.8	3.1
	Ethylglycolacetate	7.8	2.3	5.2
Alcohols	Ethanol	7.7	4.3	9.5
	iso-Propanol	7.7	3.0	8.0
	n-Butanol	7.8	2.8	7.7
Glycole ethers	Ethylglycole	7.9	4.5	7.0
	Butylglycole	7.8	2.5	6.0
	Butyldiglycole	7.8	3.4	5.2

Equation 6.59

$$\delta_d = \sum_i \delta_{d,i} \cdot \phi_i \qquad (1)$$

$$\delta_p = \sum_i \delta_{p,i} \cdot \phi_i \qquad (2)$$

$$\delta_h = \sum_i \delta_{h,i} \cdot \phi_i \qquad (3)$$

This has the consequence that, for example, two solvents that both lay outside of the solubility sphere, but whose connecting line passes through it may, when mixed with each other, be a good solvent for the polymer. This rule opens the door to finding the solubility parameters of polymers by dissolution experiments in solvent mixtures of which the solubility parameters are known. A further method relies on the use of inverse gas chromatography. Details of this technique are found in literature [72, 73].

If the solubility parameters of a resin are known, then pigment stabilization can be selectively accomplished by choosing the right solvent combinations. By adding solvents that lay outside of the solubility sphere, the adsorption onto pigment surfaces can be enhanced. In that case, it is irrelevant into which direction the system is changed, if to lesser dispersive, to lesser polar or to lesser hydrogen bonding interactions. The addition of solvents that increase the solubility of the polymer let's the adsorption layer swell, which improves pigment stabilization. The utilization of these concepts was already mentioned in Chapter 6.5.2. Tables 6.1 and 6.2 list the solubility parameters of a number of solvents and resins.

On many occasions, pigments become instable against flocculation during the "flash-off time" when the solvents leave the applied paint film. Depending upon

Table 6.2: Solubility parameters and solubility limits R_0 of some commercial paint resins

source: Shell Chemie Brochure "Löslichkeitsparameter", (= solubility parameter)", LA-08

Polymer type	Trade name	Producer	δ_d	δ_p	δ_h	R_0
Polymethyl methacrylate	Elvacite 2042	Lucite	8.60	4.72	1.94	4.20
Epoxy resin	Epicote 1001	Shell Chemie	9.95	5.88	5.61	6.00
Long oil alkyde	Plexal P65	Polyplex	9.98	1.68	2.23	6.70
Short oil alkyde	Plexal C 34	Polyplex	9.04	4.50	2.40	5.20
Polyvinylacetate	Mowilith 50	Celanese	10.23	5.51	4.72	6.70
Celluloseacetate	Cellidora A	Bayer	9.08	6.22	5.38	3.70
Saturated polyester	Desmophen 850	Bayer	10.53	7.30	6.00	8.20
Hexamethoxy methyl melamine	Cymel 300	Cytec	9.95	4.17	5.20	7.20

the evaporation numbers[19] of the solvents used, not only the total content of solvents in the coating can change, but also their composition. If at a time when the pigments are still mobile thermodynamically unfavourable solvents prevail, the steric stabilization mechanism may fail.

There is also a relationship between the Hildebrand solubility parameter δ and the enthalpic part of the Flory-Huggins interaction parameter χ_H (see Equation 6.35) [74]. If the solubility parameters of polymer (index 2) and solvent (index 1) are similar, then χ_H may be expressed as

Equation 6.60 $$\chi_H = \frac{V_1}{RT} \cdot (\delta_1 - \delta_2)^2$$

The entropic part of the interaction parameter χ_S normally lies between 0.3 and 0.4. Often a value of 0.34 is used. Then, the Flory-Huggins interaction parameter χ becomes (see Equation 6.35)

Equation 6.61 $$\chi = 0.34 + \frac{V_1}{RT} (\delta_1 - \delta_2)^2$$

If polar interactions and hydrogen bonds are to be considered, then χ can be calculated from Equation 6.62.

Equation 6.62 $$\chi = 0.34 + \frac{V_1}{RT} \left[(\delta_{d,1} - \delta_{d,2})^2 + (\delta_{p,1} - \delta_{p,2})^2 + (\delta_{h,1} - \delta_{h,2})^2 \right]$$

6.7 Adsorption of polymers on pigment surfaces

Generally, "adsorption isotherms" are used to describe the adsorption of molecules onto solid surfaces. Adsorption isotherms are graphical plots of the adsorbed amount of an adsorbate on an adsorbens as a function of the initial concentration or, in the case of gases, the partial pressure of the adsorbate. They are called "isotherms" because the adsorption values are measured while keeping the temperature constant.

Well known types of adsorption isotherms are those of *Henry*, *Langmuir* and *Freundlich* as well as the adsorption isotherm of *Brunauer*, *Emmet* and *Teller* (BET). These adsorption isotherms describe the adsorption of one type of adsorbate at any one time. The adsorption of polymer molecules from a solution is however different in the sense that at least two types of molecules, a solvent and a polymer, compete with each other for the adsorption sites. That is why this type

[19] The evaporation number states how much longer time a solvent needs to evaporate relative to a standard solvent (mostly diethylether in Europe or n-butyl acetate in the USA and Asia) at room temperature and 65 % relative humidity.

of adsorption is called "competitive adsorption". A different way of making a picture of the situation is to consider a polymer as an entity having a certain amount of solvent bound to it when it adsorbs to a surface.

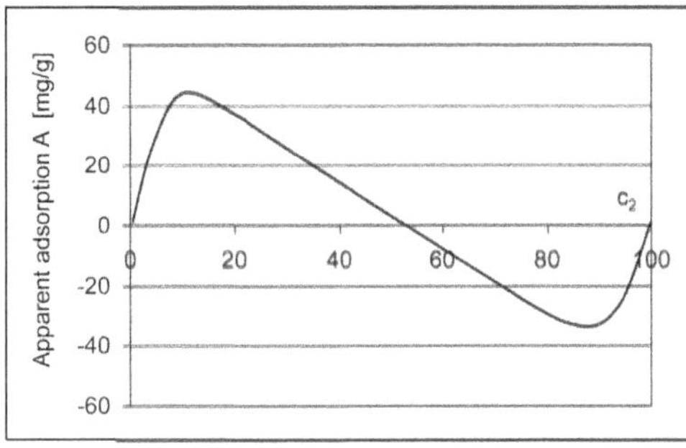

Figure 6.29: Competitive adsorption isotherm

To measure a competitive adsorption isotherm, a polymer solution of concentration c_1 (in grams polymer per 100 grams of solution) is brought into contact with a pigment at a certain weight ratio of M_1 (grams of polymer solution per gram of pigment). When the adsorption equilibrium is reached, the pigment is centrifuged and the concentration of the polymer c_2 in the supernatant solution is determined. This is done for growing polymer concentrations c_1 and the apparently adsorbed amount of polymer A is calculated in every case using the equation

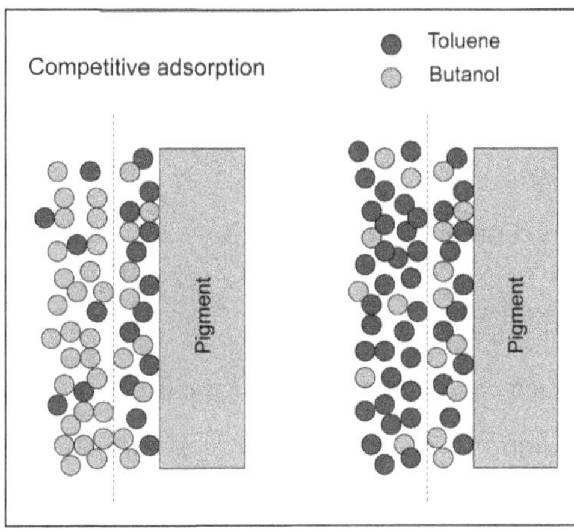

Figure 6.30: Sketch explaining the principle reason for the linear part of the competitive adsorption isotherm by a constant composition of the adsorption layer.

Equation 6.63

$$A = M_1 \cdot (c_1 - c_2)$$

A has the dimension gram of polymer per gram of pigment (and is conveniently reported as mg polymer per gram of pigment). If the apparently adsorbed polymer amount A is plotted against c_2, the concentration of polymer after adsorption, then adsorption isotherms result that start from zero, increase to positive values, go through a maximum and then decline linearly to cross the concentration axis. Since highly concentrated polymer solutions are so viscous that the pigments are not separable, the complete course of the adsorption isotherm cannot be determined. If, however, two solvents, for example toluene and butanol, are used in a competitive adsorption experiment onto pigment surfaces[20], the complete adsorption isotherm can be measured. After intersecting

[20] Personal communication of O. J. Schmitz, Paint Research Institute in Stuttgart, Germany.

with the concentration axis, the isotherm goes through a minimum and then ends at zero apparent adsorption. This is shown schematically in Figure 6.29.

According to *Rehacek* and *Schuette* [75], this type of adsorption isotherm can be explained by assuming that the composition of the adsorption layer remains constant in the region where the adsorption isotherm is linear. This is shown in Figure 6.30. The dotted line in Figure 6.30 is supposed to symbolize the boundary of the adsorption layer. If M_a is the ratio of the weight of the resin solution in the adsorption layer to the amount of pigment used, then, after adsorption (index 2) the mass balance

Equation 6.64 $$M_2 = M_1 - M_a$$

can be written. Taking c_a as the concentration of polymer in the adsorption layer, a polymer mass balance can also be defined whereby the amount of polymer that was added to the pigment is either within the adsorbed layer or in the solution surrounding the particles.

Equation 6.65 $$M_1 \cdot c_1 = M_a \cdot c_a + M_2 \cdot c_2$$

Insertion of M_2 from Equation 6.64 into 6.65 and rearranging yields

Equation 6.66 $$c_1 - c_2 = \frac{M_a c_a - M_a c_2}{M_1}$$

which can be inserted into Equation 6.53 to generate Equation 6.67

Equation 6.67 $$A = M_a \cdot c_a - M_a \cdot c_2$$

A plot of the apparent adsorption A against the concentration of the supernatant solution after adsorption c_2 yields a straight line from whose slope the amount of adsorbed polymer may be taken. With this value and the intercept with the y-axis, the concentration of the polymer in the adsorbed layer is found.

Rehacek studied the adsorption of alkyd resins on various inorganic pigments [76]. A paper investigating the influence of anchor groups on the adsorption of polymers on differently surface treated titanium dioxide pigments is from Idogawa et al. [77].

Typical compositions of adsorption layers are reported in Table 6.3.

Table 6.3: Typical compositions of adsorption layers of alkyd resins on inorganic surface

Adsorbed resin solution per m² pigment surface	approx. 3 mg
Concentration of adsorbed layer	25 to 40 %
Adsorbed layer thickness	7 to 15 nm

6.8 Let-down

In order to disperse cost-efficiently, mill bases are normally formulated with a larger pigment concentrations than present in the final recipe. Therefore, after dispersion, the mill bases must be completed by adding the other ingredients to come to the aspired paint composition. This is called the "let-down". However, this has to be done very carefully in order to avoid that the state of dispersion is not destroyed by flocculation again. During the let-down, concentration gradients can occur which may detrimentally change the structure and composition of the adsorption layers.

If the resin concentration in the medium used for completion is too high, then, due to the osmotic pressure, solvent may be drawn out of the adsorption layer. The adsorption layer will then be compressed so that the onset of repulsion by steric stabilization will be at closer pigment particle separation distances where the van der Waals attraction is already large. Furthermore, when adsorption layers are compressed by the extraction of solvent, less osmotic pressure of the solvent is generated when the pigment particles approach each other so that the mechanism of steric stabilization may not come into effect at all. This particular case is called **"pigment shock"**.

The **"resin shock"** can occur when too much thermodynamically poor solvent is introduced in the let-down. Thereby the adsorption layer can also be compressed since the loops and tails are dissolved less well. This has the same consequence as in the case of a pigment shock.

A **"solvent shock"** can happen when too much thermodynamically favourable solvent is added in the let-down. In extreme cases it can happen that adsorbed layers are disconnected from the pigment surface. Then, the pigments are unprotected and flocculate. Table 6.4 lists all the possible reasons for failure during the let-down [78].

In order to avoid all of these phenomena, an adequate formulation philosophy should be followed. The mill base should best have all the major ingredients of the final formulation to a certain degree. Especially the solvent composition in

Table 6.4: Reasons for failures during the let down

	Let down paste	Cause
Pigment shock	Too much resin	Resin solution of the let down extracts solvent out of the mill base
Resin shock	With thermodynamically poor solvent	Coagulation of the resin in the mill base
Solvent shock	Pure solvent or very diluted resin solution	Detachment of adsorbed polymer layers

the mill base should be similar to that of the final formulation. On the other hand, it is important that the let-down is done while stirring sufficiently to avoid large concentration gradients. On many occasions the order in which the components are mixed plays an important role. This is explained by different concentration situations. It can also happen that results are different if either component A is added to component B or vice versa. When developing formulations, the production procedure should be accounted for from the beginning on.

6.9 Flocculation stabilization by rheology control

A further method to prevent flocculation is to introduce a yield point (see Chapter 3.5) into the formulations. A yield point characterizes the situation when in the resting liquid, the elastic components of deformation are larger than the viscous parts. A minimum stress is necessary to let the medium flow at all. An example for such a system is tooth paste. When it is pressed out of the tube, it flows like a liquid. Yet, on the bristles of the tooth brush it rests like a solid body. In a system having a yield point, the Brownian motion is suppressed. If the yield point in a paint is large enough, pigments can neither flocculate nor settle. In water based formulations, "inorganic thickeners" are often used to create a yield point. These are layer structured silicates ("phyllosilicates") like bentonite, naturally occurring aluminium silicates or laponite, a synthetic magnesium silicate. They consist of silicate anion layers that, in the crystals, alternate with layers of the respective cations. When put into water, the sides of the particle plates obtain a positive charge due to the cations there, whereas the planes of the plates are charged negatively. They therefore form chart house structures (see Figure 6.31) in which the positive sides of the plates arrange with the negative planes by electrostatic interaction. This generates a resistance to flow. The smaller pigment particles are trapped in between the card house structure. Other thickeners for aqueous media are high molecular weight cellulose ethers and polyacrylates. Mainly polyurethane based associative thickeners have hydrophobic as well as hydrophilic moieties within their molecular architecture that can associate either with themselves or with other paint ingredients, thereby creating a yield point. For solvent based systems, a variety of polymeric thickeners (organic thickeners) of different chemical nature are available.

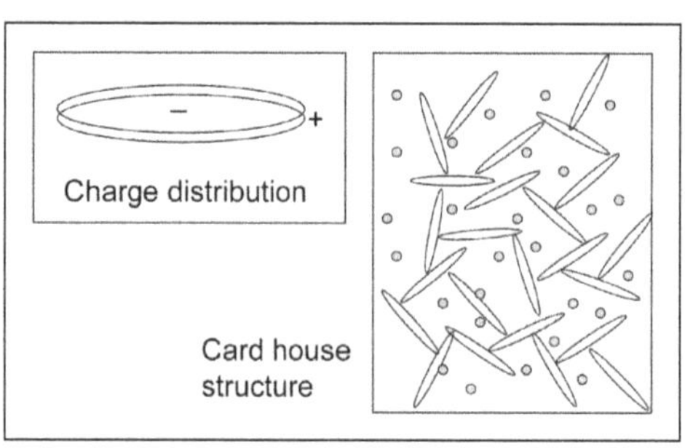

Figure 6.31: Schematic representation of pigment stabilization against flocculation by layered silcates that create a yield point due to the formation of a chart house structure

6.10 Literature

[1] M. von Smoluchowski, Zeitschr. Phys. Chem. 92 (1917) 129
[2] T. M. Herrington, B. R. Midmore, Powder Technology, 65 (1991) 251–256
[3] V.E. Proskurina, V.A. Myagchenkov, Russ. J. Appl. Chem. 79 (2006) 301–3005
[4] Y. Kawashima, T. Handa, H. Takeuchi, Y. II, H. Takenaka, Powder Technology 46 (1986) 61–66
[5] J. A. Maroto, F. J. de las Nieves, Colloid Polymer Sci. 275 (1997) 1148
[6] O. Soehnel, J. W. Mulin, AIChE Symposium Series 87 (1991) 182–190
[7] N. Fuchs, Zeitschrift für Physik, 89 (1934) 736–743
[8] R. L. Soulsby, "Dynamics of Marine Sands; A Manual for Practical Applications", Thomas Telford, London, 1987
[9] N. S. Cheng, J. Hydraul. Eng. 123 (1997) 149–152
[10] J. A. Jiminez, O. S. Madsen, Journal of Waterway, Port, Coastal and Ocean Engineering, March/April 2003, 70–78
[11] E. J. W. Verwey, J. Th. G. Overbeek, Theory of the Stability of Lyophobic Colloids, Elsevier, Amsterdam, 1948;
[12] M. J. Vold, J. Colloid Sci., 16 (1961) 1-12
[13] T. Crowl, M. A. Malati, Discussions Faraday Soc., 42 (1966) 301–312
[14] B. V Derjaguin, Trans. Faraday Soc. 36 (1940) 730
[15] B. Derjaguin, L. Landau, "Theory of the stability of strongly charged lyophobic sols and of the adhesion of strongly charged particles in solutions of electrolytes", Acta Physico Chemica URSS 14 (1941) 633
[16] M. R. Böhmer, L.K. Koopal, Langmuir, 8 (1992) 2649–2659, und 8 (1992) 2660–2665
[17] G. Lagaly, Colloids, in "Ullmanns Encyclopedia of Industrial Chemistry", 5th Ed., VCH, Weinheim, 1986, Vol. A7, p. 341-367
[18] J. Winkler, "Titanium Dioxide", Vincentz Network, Hannover, 2003, p. 74
[19] Personal communication of Dr. Günther Michael, Evonik Degussa
[20] T. Rentschler, farbe +lack 106 (2000) 62
[21] C.-P. Huang, "The Surface Acidity of Hydrous Solids" in "Adsorption of Inorganics at Solid-Liquid Interfaces", M. A. Anderson and A. J. Rubin, Editors, Ann Arbor Science, Ann Arbor, MI, 1981, Chapter 5
[22] J. Winkler, farbe + lack 97 (1991) 859–866
[23] K. Kato, K. Kanda, H. Kondo, K. Esumi, K. Meguro, Bull. Chem. Soc. Jpn, 60 (1987) 3379–3383
[24] Unpublished research results of the author
[25] zitiert in D.C. Grahame, Chem. Rev. 41 (1947) 441–501
[26] H. Ohshima in "Colloid Stability; The Role of Surface Forces – Part 1, Volume 1", T. F. Tadros Ed., Wiley-VCH Verlag, 2007, S. 50–52
[27] R. W. O'Brian, J. Fluid Mech. 190 (1988) 71-86
[28] J. Winkler, Journal of Coatings Technology, 65, No. 822 (1993) 29–36
[29] J. Winkler, Fatipec Congress, 15th –19th May, Budapest, Congress Book, Volume 2, p. 284-296
[30] S. H. Behrens, M. Borkovec, Journal of Chemical Physics 111 (1999) 382–385

[31] B. V. Zhmud, L. Bergström, "Charge Regulation at the Surface of a Porous Solid", Surfactant Science Series 78, J. A. Schwarz, C. I. Contescu Ed., Marcel Dekker (1999) 567–592
[32] J. Lyklema, Advan. Colloid Interface Sci. 2 (1968) 65–114
[33] H. Ohshima in "Colloid Stability; The Role of Surface Forces – Part 1, Volume 1", T. F. Tadros Ed., Wiley-VCH Verlag, 2007, S. 50–52
[34] D. H. Napper, Kollid Zeitschrift und Zeitschrift für Polymere, 234 (1969) 1149–1151
[35] E. L. Mackor, J. Colloid Sci. 6 (1951) 492–495
[36] E. L. Mackor, J. H. van der Waals, J. Colloid Sci. 7 (1952) 535
[37] E. J. Clayfiel, E. C. Lumb, J. Colloid Interface Sci. 22 (1966) 269–284
[38] E. J. Clayfiel, E. C. Lumb, J. Colloid Interface Sci. 22 (1966) 285–293
[39] E. W. Fischer, Kolloid Zeitschrift 160, Nr. 2 (1958) 120–141
[40] D. H. Napper, Transactions Faraday Soc. 64 (1968) 1701
[41] D. H. Napper, Journal of Colloid and Interface Science 32, No. 1 (1970) 106–114
[42] D. H. Napper, Journal of Colloid and Interface Science 58, No. 2 (1977) 390–407
[43] P. C. Hiemenz, Polymer Chemistry, Marcel Dekker, New York (1984) 43 ff.
[44] P. J. Flory, J. Chem. Phys. 9 (1941) 660
[45] P. J. Flory, J. Chem. Phys. 10 (1942) 51
[46] M. L. Huggins, J. Chem. Phys. 9 (1941) 440
[47] M. L. Huggins, J Phys. Chem. 46 (1942) 151
[48] M. L. Huggins, Ann. N. Y. Acad. Sci. 43 (1942) 1
[49] M. L. Huggins, J. Am. Chem. Soc. 64 (1942) 1712
[50] P. J. Flory, W. R. Krigbaum, J. Chem. Phys. 18 (1950) 1086
[51] E. Jenckel, B. Rumbach, Z. Elektrochem. 55 (1951) 612
[52] G. R. Joppien, Die Makromolekulare Chemie 175 (1974) 1931–1954
[53] G. R. Joppien, Die Makromolekulare Chemie 176 (1975) 1129–1149
[54] R. E. Felter, L. N. Ray, Journal of Colloid and Interface Science 32, No. 2 (1970) 349–360
[55] R. E. Felter, E. S. Moyer, L. N. Ray, J. Polymer Science Vol. 7 (1969) 529–533
[56] C. van der Linden, R. van Leemput, Journal of Colloid and Interface Science 67, No. 1 (1978) 63–69
[57] G. S. Sadakne, J. L. White, Journal of Applied Polymer Science 17 (1973) 453–469
[58] M. A. Cohen Stuart, J. M. H. M. Scheutjens, G. F. Fleer, Journal of Polymer Science, Polymer Physics Edition, Vol. 18 (1980) 559–573
[59] J. Winkler, L. Dulog, farbe und lack 89 (1983) 236–242
[60] J. Winkler, W.-R. Karl, farbe + lack 100 (1994) 171–176
[61] J. H. Hildebrand, R. L. Scott, "Solubility of Non-electrolytes", Reinhold, New York, 1949
[62] G. Scatchard, Chem. Rev. 8 (1931) 321–333
[63] C. M. Hansen, Journal of Paint Technology, 39, No. 505 (1967) 104–117
[64] C. M. Hansen, Journal of Paint Technology, 39, No. 511 (1967) 505–510
[65] C. M. Hansen, K. Skaarup, Journal of Paint Technology, 39, No. 511 (1967) 511–514
[66] A. Bondi, D. J. Simken, A. I. Ch. E. Journal 3 (1957) 473
[67] D. M. Koehnen, C. A. Smolders, J. Appl. Polymer Science 19 (1975) 1163

[68] J. H. Sewell, R. A. E. Technical Report No. 66185, June 1966
[69] C. F. Böttcher, "Theory of Electric Polarization", Elsevier, New York, 1952, Chapter 5
[70] J. V. Koleske, Paint and Coating Testing Manual, 14th Edition of the Gardner-Sward Handbook, ASTM International, West Conshohocken, 1995, Chapter 35
[71] C. M. Hansen, A. Beerbower, "Solubility Parameters", Kirk Othmer Encyclopedia of Chemical Technology, Supplement Volume 2, 2nd Edition, A. Standen, Ed., Interscience, New York, 1971, pp. 889–910
[72] G. Dipaola-Baranyi, J. E. Guillet, Macromolecules 11 (1978) 147–154
[73] J. E. G. Lipson, J. E. Guillet, Journal of Polymer Science: Physics Edition, 19 (1981) 1199–1209
[74] E. A. Grulke "Solubility Parameter Values" in "Polymer Handbook", J. Brandrup, E. H. Immergut, E. A. Grulke, A. Abe, D. R. Bloch Ed., 2005, John Wiley & Sons, NY
[75] K. Rehacek, H. Schütte, Plaste und Kautschuk 16 (1969) 773
[76] K. Rehacek, Farbe und Lack 16 (1970) 656
[77] H. Idogawa, O. Shimizu, K. Esumi, Journal of Coatings Tecnology 65, No. 823 (1993) 67–73
[78] H. Sander, farbe und lack 77 (1971) 891–892

Appendix

Calculation of the polar and dispersive contributions of the free surface energy of solids according to Wu

The set of Equations 3.21 is given by:

$$(b_1 + c_1 - a_1) \cdot \gamma_s^d \gamma_s^p + c_1(b_1 - a_1) \cdot \gamma_s^d + b_1(c_1 - a_1) \cdot \gamma_s^p - a_1 \cdot b_1 \cdot c_1 = 0$$

$$(b_2 + c_2 - a_2) \cdot \gamma_s^d \gamma_s^p + c_2(b_2 - a_2) \cdot \gamma_s^d + b_2(c_2 - a_2) \cdot \gamma_s^p - a_2 \cdot b_2 \cdot c_2 = 0$$

in which

$$a_1 = \frac{1}{4}\gamma_1(1 + \cos\theta_1) \qquad b_1 = \gamma_1^d \qquad c_1 = \gamma_1^p \qquad \theta_1 = \text{contact angle of liquid 1}$$

$$a_2 = \frac{1}{4}\gamma_2(1 + \cos\theta_2) \qquad b_2 = \gamma_2^d \qquad c_2 = \gamma_2^p \qquad \theta_2 = \text{contact angle of liquid 2}$$

So, contact angle measurements are needed using two liquids for which the polar and the dispersive contributions to the surface tension have to be known.

To calculate the polar and dispersive contributions of the free surface energy of a solid from contact angle measurements from different liquids, it is best to first carry out the following substitutions in equation 3.21 in order to have a better overview.

$$k_1 = (b_1 + c_1 - a_1) \qquad l_1 = c_1(b_1 - a_1) \qquad m_1 = b_1(c_1 - a_1) \qquad n_1 = a_1 b_1 c_1$$

$$k_2 = (b_2 + c_2 - a_2) \qquad l_2 = c_2(b_2 - a_2) \qquad m_2 = b_2(c_2 - a_2) \qquad n_2 = a_2 b_2 c_2$$

Then Equation 3.21 then simplifies to:

$$I. \quad k_1 \cdot \gamma_s^d \cdot \gamma_s^p + l_1 \cdot \gamma_s^d + m_1 \gamma_s^p - n_1 = 0$$

$$II. \quad k_2 \cdot \gamma_s^d \cdot \gamma_s^p + l_2 \cdot \gamma_s^d + m_2 \gamma_s^p - n_2 = 0$$

Jochen Winkler: Dispersing Pigments and Fillers

ISBN 978-3-86630-852-7

By rearranging Equation II, the polar contribution of the free surface energy of the solid is isolated:

$$III.\quad \gamma_s^p = \frac{n_2 - l_2 \cdot \gamma_s^d}{m_2 + k_2 \cdot \gamma_s^d}$$

Replacing γ_s^p in equation I by the right hand side of Equation III yields:

$$k_1 \cdot \gamma_s^d \cdot \frac{n_2 - l_2 \cdot \gamma_s^d}{m_2 + k_2 \cdot \gamma_s^d} + l_1 \cdot \gamma_s^d + m_1 \cdot \frac{n_2 - l_2 \cdot \gamma_s^d}{m_2 + k_2 \cdot \gamma_s^d} - n_1 = 0$$

$$k_1 \cdot \gamma_s^d(n_2 - l_2 \cdot \gamma_s^d) + l_1 \cdot \gamma_s^d(m_2 + k_2\gamma_s^d) + m_1(n_2 - l_2 \cdot \gamma_s^d) - n_1(m_2 + k_2\gamma_s^d) = 0$$

$$n_2k_1\gamma_s^d - l_2k_1(\gamma_s^d)^2 + m_2l_1\gamma_s^d + l_1k_2(\gamma_s^d)^2 + m_1n_2 - m_1l_2\gamma_s^d - n_1m - n_1k_2\gamma_s^d = 0$$

$$IV.\quad \underbrace{(\gamma_s^d)^2(l_1 \cdot k_2 - l_2 \cdot k_1)}_{o} + \underbrace{\gamma_s^d(n_2 \cdot k_1 + m_2 \cdot l_1 - m_1 \cdot l_2 - n_1k_2)}_{p} + \underbrace{(m_1n_2 - n_1m_2)}_{q} = 0$$

Equation IV is a quadratic equation with the solutions:

$$V.\quad \gamma_{s,1,2}^d = \frac{-p \pm \sqrt{p^2 - 4 \cdot o \cdot q}}{2 \cdot o}$$

Re-substitution of all variables leads to the following expressions for o, p and q:

$$o = c_1(b_1 - a_1) \cdot (b_2 + c_2 - a_2) - c_2(b_2 - a_2) \cdot (b_1 + c_1 - a_1)$$

$$p = a_2b_2c_2(b_1 + c_1 - a_1) + b_2c_1(c_2 - a_2)(b_1 - a_1) - b_1c_2(c_1 - a_1)(b_2 - a_2) - a_1b_1c_1(b_2 + c_2 - a_2)$$

$$q = a_1b_1^2c_1(c_1 - a_1) - a_2b_2^2c_2(c_2 - a_2)$$

Using these values, γ_s^d can be calculated from Equation V and finally γ_s^p can be calculated either from Equation I or from Equation II, respectively.

Author

Jochen Winkler studied chemistry at the University of Stuttgart, Germany. From 1980 until 1984 he worked with the German Paint Research Institute (Forschungsinstitut für Pigmenten und Lacke, FPL) in Stuttgart, where he studied pigment flocculation and did work on the energy balancing of dispersion machinery. He joined Sachtleben Chemie GmbH in Duisburg, Germany, as member of their technical service laboratory in 1985, but soon moved on to the R&D labs. In 1992 he became responsible for R&D at Sachtleben. In the year 2009 he joined Hemmelrath GmbH, a German automobile paint producer, where he managed the paint development laboratories. The author joined Crenox GmbH in Krefeld, Germany, in 2010, where he is currently responsible for Corporate Development. Jochen Winkler has been holding classes on colloidal chemistry at the University of Stuttgart for many years and was awarded with an honorary professorship from there.

Jochen Winkler: Dispersing Pigments and Fillers

ISBN 978-3-86630-852-7

Index

A

B

C

D

E

F

G

H

I

P

R

S

T

V

W

X

Y

Z

9 783866 308527

www.ingramcontent.com/pod-product-compliance
Lightning Source LLC
LaVergne TN
LVHW020638300926
849161LV00004B/745